Waves in Fluids and Solids

Waves in Fluids and Solids

Editor

Fred Gajos

scitus
academics

Waves in Fluids and Solids
Edited by **Fred Gajos**

ISBN: 978-1-68117-662-8
Library of Congress Control Number: 2016946529

616, Corporate Way, Suite 2, 4766,
Valley Cottage, NY 10989

www.scitusacademics.com

Printed in United States of America on Acid Free Paper

Preface

The propagation of waves in fluids and solids is one of the ways in which waves travel. As regards the path of oscillation in relation to the direction of propagation, we can distinguish between longitudinal wave and transverse wave. Propagation in a vacuum and in a material medium can occur in electromagnetic waves. Other types of waves cannot propagate through a vacuum and you must have a transmission medium. Mechanical waves are waves that propagate at a wave speed that is determined by the inertial and elastic properties of a material medium (solid, fluid or gas).

This book is intended to emphasize and compare the various areas of interest, including the necessary scientific calculations, constituent theories for different materials and products, polarized and shock waves, and second sound in solids at low temperatures. This book also focuses on acoustic waves in heterogeneous media in elastic disturbances and fluid media. Various systems such as layered media, phonic crystals, interface waves, granular materials, crystalline systems, piezoelectric substrates, solutions, soft media and acoustic levitation are analyzed in this book.

Table Of Contents

Chapter 1. **Computational Study of the Effect of Cortical Porosity on Ultrasound Wave Propagation in Healthy and Osteoporotic Long Bones** 1

Chapter 2. **Theoretical Analysis of SAW Propagation in 3C-SiC/c-AlN** 49

Chapter 3. **Dynamic Characterization of Cohesive Material Based on Wave Velocity Measurements** 77

Chapter 4. **Unidirectional Wave Propagation in Low-Symmetric Colloidal Photonic-Crystal Heterostructures** 125

Chapter 5. **Analytical Study of the Propagation of Fast Longitudinal Modes along wz-BN/AlN Thin Acoustic Waveguides** 143

Chapter 6. **A Scalar Acoustic Equation for Gases, Liquids, and Solids, Including Viscoelastic Media** 165

Chapter 7. **Influence of the Chemical Composition of Completion Fluids on the Propagation of Electromagnetic Waves within Oil Wells** 179

Chapter 8. **Analytical Solution for Acoustic Waves Propagation in Fluids 197**

Chapter 9. **Antiplane elastic wave propagation in pre-stressed periodic structures; tuning, band gap switching and invariance 209**

Chapter 10. **A model of acoustic absorption in fluids based on a continuous distribution of relaxation times 245**

Index 293

CHAPTER 1

Computational Study of the Effect of Cortical Porosity on Ultrasound Wave Propagation in Healthy and Osteoporotic Long Bones

Vassiliki T. Potsika [1], Konstantinos N. Grivas [2], Theodoros Gortsas [2], Gianluca Iori [3], Vasilios C. Protopappas [1], Kay Raum [3], Demosthenes Polyzos [2] and Dimitrios I. Fotiadis [1,4,*]

[1] Unit of Medical Technology and Intelligent Information Systems, Department of Materials Science and Engineering, University of Ioannina, GR 45110 Ioannina, Greece
[2] Department of Mechanical Engineering and Aeronautics, University of Patras, GR 26500 Patras, Greece
[3] Berlin-Brandenburg School for Regenerative Therapies, Charité-Universitätsmedizin Berlin, Augustenburger Platz 1, 13353 Berlin, Germany
[4] Foundation for Research and Technology–Hellas, Institute of Molecular Biology and Biotechnology, Department of Biomedical Research, GR 45110 Ioannina, Greece

ABSTRACT

Computational studies on the evaluation of bone status in cases of pathologies have gained significant interest in recent years. This work presents a parametric and systematic numerical study on ultrasound propagation in cortical bone models to investigate the effect of changes in cortical porosity and the occurrence of large basic multicellular units, simply called non-refilled resorption lacunae (RL), on the velocity of the first arriving signal (FAS). Two-dimensional geometries of cortical bone are established for various microstructural models mimicking normal and pathological tissue states. Emphasis is given on the detection of RL formation which may provoke the thinning of the cortical cortex and the increase of porosity at a later stage of the disease. The central excitation frequencies 0.5 and 1 MHz are examined. The proposed configuration consists of one point source and multiple successive receivers in order to calculate the FAS velocity in small propagation paths (local velocity) and derive a variation profile along the cortical surface. It was shown that: (a) the local FAS velocity can capture porosity changes including the occurrence of RL with different number, size and depth of formation; and (b) the excitation frequency 0.5 MHz is more sensitive for the assessment of cortical microstructure.

KEYWORDS

ultrasound; bone modeling; osteoporosis; cortical porosity; basic multicellular units

1. INTRODUCTION

Osteoporosis is a bone pathology characterized by low bone mass and structural deterioration of bone tissue leading to bone fragility, thinning of the cortical cortex, reduced bone strength and an increased susceptibility to fractures. Nowadays, the ageing of the population has led to greater prevalence of osteoporosis which is responsible for approximately 2 million fractures which occur annually and may lead to diminished quality of life, disability and even death with a mortality rate 36% depending on the age [1,2]. In clinical practice, the golden standard for the assessment of the disease is dual energy X-ray absorptiometry scan. However, X-ray-based imaging techniques are expensive methods and require radiation exposure. Therefore, the development of non-destructive and non-radiating methods, appropriate to assess the mechanical and structural properties of bone, is crucial.

To this end, several research groups worldwide have investigated the diagnostic and monitoring role of quantitative ultrasound in osteoporotic and fractured bones using experimental, numerical and theoretical means [3,4,5,6,7,8,9]. The research interest in numerical and theoretical methods has allowed the study of phenomena which cannot be observed via experimental measurements extending thus our understanding on the underlying ultrasound propagation mechanisms and the interaction with biological tissues [10]. Specifically, as ultrasound propagates in composite media such as osseous tissues attenuation mechanisms evolve including absorption and scattering phenomena. In soft tissues the viscous forces between neighboring particles moving with different velocities are major sources of acoustic wave absorption [1]. Scattering is induced by the porous nature of cortical bone and the incident energy is partially transferred to the scattered fields. In soft tissues like blood and marrow, ultrasound bulk shear waves are usually ignored as shear waves are highly attenuated at ultrasonic frequencies, while in hard osseous tissues both compression and shear waves are considered.

In addition, the availability of advanced computer-aided computational tools in combination with imaging data derived from micro-computed tomography (μCT) and scanning acoustic microscopy (SAM) have accelerated the evolution of the numerical research [2,3,7,8]. The axial and through transmission techniques have been mainly used to estimate the first arriving signal (FAS) velocity and attenuation as well as to study the propagation of guided waves. In the literature, different thresholds for the detection of the FAS and methodologies to extract guided wave features have been proposed to interpret ultrasound propagation phenomena occurring not only at the periosteal region, but also at deeper bone layers [4,5,6,7,11,12]. More recently, ultrasound reflection and backscattering parameters have been measured to investigate bone microstructure [13,14].

Concerning the first two-dimensional (2D) computational studies, the cortical bone was modeled as a linear elastic homogeneous plate and ultrasound simulations were performed for different plate thicknesses to investigate the thinning of the cortical cortex due to osteoporosis [15,16]. In [17], X-ray computed tomography reconstructions were derived from the human radius in order to develop more realistic 3D computational models of cortical bone. It was shown that for plate thicknesses larger than the wavelength the FAS corresponds to a lateral wave which propagates at the bulk longitudinal velocity of bone, while for very thin plates the FAS wave propagates as the lowest-order symmetric plate mode. More recent 2D and 3D computational studies [2,8,14,17,18,19] take into consideration cortical microstructure, porosity and anisotropy. In [18], SAM images of human femoral neck were used to develop numerical models which account for the sample's overall shape, microstructure, cortical porosity, heterogeneous matrix elasticity and density. It was found that the FAS velocity is not influenced by trabecular bone properties or by the heterogeneities of the cortical bone mineralized matrix. On the other hand, the FAS was sensitive to variations in cortical porosity. In addition, in [2], SAM human femoral neck cross-sections were used to develop multivariate models for the prediction of pore size, porosity, and cortical thickness. It was shown that the FAS velocity decreases with increasing porosity, while an increase of the cortical thickness and pores' diameter leads to an increase in FAS velocity. It was also reported that cortical porosity has the strongest effect on the prediction of the FAS velocity. In addition, a 3D micromechanical model consisting of an anisotropic matrix pervaded by cylindrical pores was developed in [19]. It was found that, for the elderly population, the elastic properties of the mineralized matrix do not undergo large variations among different samples, while changes in the intracortical porosity account for most of the variations of mesoscopic elasticity [19].

In this work, we present a 2D computational study of ultrasonic propagation on healthy and osteoporotic cortical bone to examine the effect of porosity on the FAS velocity. We account for the occurrence of large basic multicellular units, simply called non-refilled resorption lacunae (RL), as the predominance of resorption causes an increase in Haversian canal size leading to the increase of cortical porosity and fragility [14,20]. Initially, various microstructural models are established, mimicking normal and pathological tissue states with porosity of 5%, 10% and 16%. Then, numerical models are presented to investigate the impact of changes in the number, size, position and depth of the RL on the FAS. The central excitation frequencies 0.5 and 1 MHz are examined. To our knowledge, this is the first systematic and parametric numerical study which evaluates the FAS velocity variation along cortical bone derived from small, serial propagation paths for different porosities and frequencies, including the occurrence of pores with different sizes.

2. VALIDATION OF THE NUMERICAL METHOD: A BENCHMARK PROBLEM

The most popular numerical methods for ultrasonic evaluation of cortical bone are the Finite Difference, Finite Element and Boundary Element methods which solve partial differential equations. According to [21], for time domain modeling of broadband or high-frequency waves these methods can become cumbersome and slow due to the requirements for many grid points per wavelength and small time-steps to minimize unwanted numerical dispersion. In order to validate the finite-difference time-domain (FDTD) code that is used in this work (SimSonic, LIP, Paris, France [22,23]), a benchmark problem was established. We aim to compare the outcome of the FDTD method with a MATLAB (Version, The Mathworks, Inc., Natick, MA, USA) toolbox which combines pseudo-spectral and k-space methods, so the spatial gradients are calculated by using Fast Fourier Transforms rather than by using a finite difference stencil (k-Wave, [21]). The computational model ofFigure 1a was used corresponding to an intact cortical plate with thickness of 4 mm and length 40 mm surrounded by water. The material properties of bone and water are presented in Table 1. The grid size was set to 20 μm and the wave propagation simulation time was 10 μs. A point source and receiver were used to calculate the velocity at a distance of 12.5 mm between the transducers. The transducers were placed directly onto the cortical surface. A 1-MHz Gaussian pulse was used as the excitation signal. According to the results of Figure 2, the two waveforms have the same time of arrival of the FAS as well as the same qualititative behavior over time enhancing the accuracy of the FDTD method.

In order to provide quantitative information for the differences of the two waveforms, a convergence study is conducted for the grid sizes of 50, 30 and 20 μm. The methodology of waveform rectification was used followed by numerical integration via the trapezoidal method. A full-wave rectifier converts the examined waveform to a waveform of constant polarity (positive or negative). Specifically, the excitation signal was rectified as well as the received waveforms from k-Wave and Simsonic. Numerical integration was applied using the trapezoidal method to derive the emitted and received signal area. Next, we calculated the ratio of the emitted signal area to the received waveforms' area, which is a quantitative indicator reflecting the whole area of interest. If we denote the calculated ratio with E_{sim} for Simsonic and E_{kw} for k-Wave, the corresponding values were: (a) $E_{kw} = 1166$, $E_{sim} = 1084$ ($\text{difference}_{50\mu m} = 82$) for 50 μm, (b) $E_{kw} = 1935$, $E_{sim} = 1921$ ($\text{difference}_{30\mu m}=14$) for 30 μm and (c) $E_{kw} = 2924$, $E_{sim} = 2919$ ($\text{difference}_{20\mu m} = 5$) for 20 μm.

3. NUMERICAL EVALUATION OF CORTICAL POROSITY USING ULTRASONIC TECHNIQUES

3.1. Model Geometry

In this section, we present the structural features of the computational models representing different cases of cortical porosity and pores' sizes as well as different number and distributions of RL. The structural and material properties were derived from [14], in which the Haversian canals of normal size were differentiated from the RLs to distinguish healthy bones (with no or few RL) from degenerated ones. As can be seen in Figure 1, the cortical bone was modeled as a 2D plate (length 40 mm, width 4 mm) surrounded by water (upper surface 4 mm, lower surface 2 mm).

Two sets of numerical simulations were performed in order to evaluate: (a) porosity changes from 0% to 16% covering the whole region of cortical bone (Series I_Po0-16), (b) the occurrence of a single or a cluster of RL in small propagation paths (Series II_RL). The selection of the porosities of Series I_Po0-16, and the sizes of the RL were based on previous studies investigating the interaction of ultrasound with cortical porosity [2,5,10,14]. However, it should be emphasized that the evaluation of cortical porosity depends on several parameters such as age, gender and region of interest [5]. The first set is presented in Figure 1, in which cortical bone was modeled both as a homogeneous and nonhomogeneous medium to account for the cases of: (a) intact bone (Figure 1a); (b) porosity 5% (Figure 1b); (c) porosity 10% (Figure 1c); and (d) porosity 16% (Figure 1d–f). In Table 2, we present the exact structural features concerning the porosity, number and dimensions of the normal and larger pores. For each microstructural scenario of Table 2, a total of three numerical models were established and the mean local FAS velocity values were calculated as well as the standard error.

Circular scatterers were considered in the transverse direction and the selected radii of normal pores and RL are presented in Table 2. Given the porosity and the dimensions of the plate and scatterers, the number of the pores can be calculated. An algorithm was developed to generate the osseous and soft tissues as greyscale values and the "porosity maps" with a random distribution of the pores. The scatterers were not in contact and did not merge with each other as well as with the cortical boundaries.

Figure 3 illustrates cortical segments of Series II_RL in which the cortex of a long bone was modeled initially as a 2D homogeneous plate including the presence of 1 RL at different lengths, depths and diameters (Figure 3a–e). Then, in Figure 3f–h, we account for cortical microstructure to simulate the case of normal remodeling (Table 2: cases B_Po5_RL1, B_Po5_RL3, B_Po5_RL5) including a single or a cluster of RL in the center of the plate.

3.2. Material Properties

The cortical bone was modeled as an isotropic, linear elastic medium. We assumed that the soft tissues surrounding the cortical plate as well as the circular pores are composed of water. Table 1 summarizes the material properties assigned to the cortical bone, scatterers and soft tissues which were derived from [14].

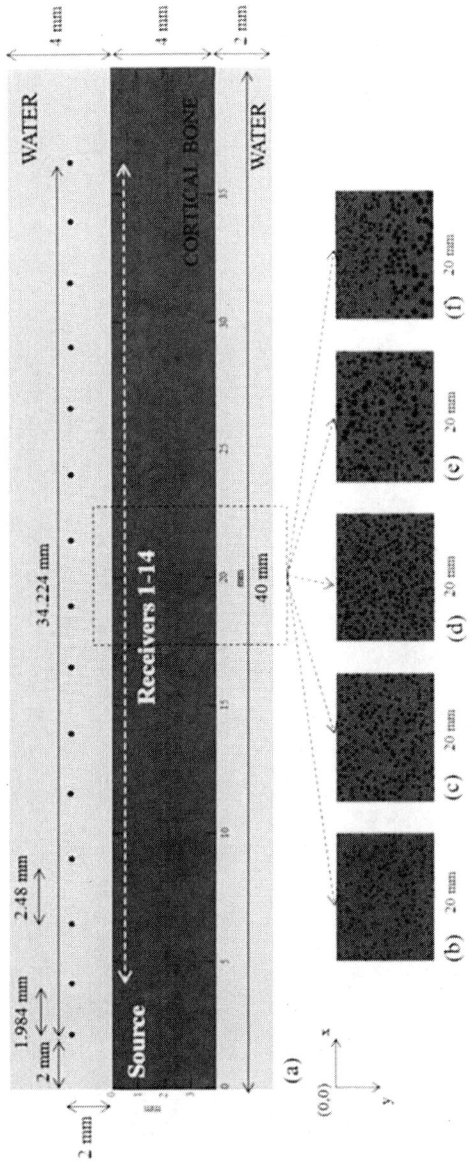

Figure 1. Numerical models of cortical bone corresponding to the first set of simulations (Series I_Po0-16) and Table 2, namely: (**a**) B_Hom; (**b**) B_Po5; (**c**) B_Po10; (**d**) B_Po16; (**e**) B_Po16_RL; and (**f**) B_Po16_Gradual. For each geometry, only one of the three random porosity distributions is depicted in the form of 5 mm cortical segments derived from the original 40mm × 4mm plates. The ultrasound configuration is also presented in (**a**).

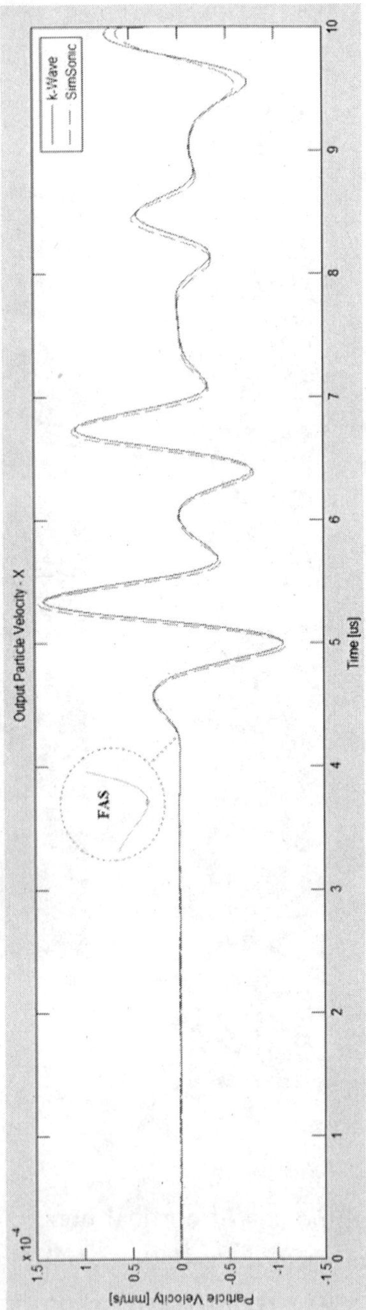

Figure 2. Particle velocity waveforms derived from two different numerical tools, Simsonic and k-Wave, for grid size 20 μm. The threshold for the detection of the FAS is also depicted.

Table 1. Elastic coefficients and density used for the simulations [14].

Propagation Medium	C_{11} (GPa)	C_{12} (GPa)	C_{66} (GPa)	P (g/cm^3)
Water	2.25	2.25	0	1.00
Bone	28.71	10.67	9.02	1.85

Table 2. Examined cases in the tangential direction for the second set of the simulations.

	Description	Porosity (%)	No. of Pores	Pores' Radius (μm)	No. of RL/RLs	Radius of RL/RLs (μm)
B_Hom	Homogeneous bone	0	0	–	–	–
B_Po5	Porous bone, normal pores	5	1592	40	–	–
B_Po5_RL1	Porous bone, normal pores and 1 RL	5	1592	40	1	115
B_Po5_RL3	Porous bone, normal pores and 3 RLs	5	1592	40	3	115
B_Po5_RL5	Porous bone, normal pores and 5 RLs	5	1592	40	5	115
B_Po10	Porous bone, normal pores	10	1592	60	–	–
B_Po16	Porous bone, normal pores	16	2263	60	–	–
B_Po16_RL	Porous bone, normal pores and RLs	16	1400	60	235	115
B_Po16_Gradual	Gradual distribution of the pores, normal pores and RLs	16	1775	40, 60, 80	192	115

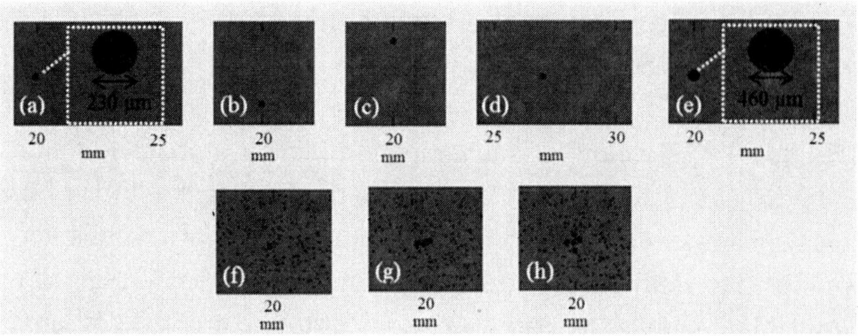

Figure 3. Numerical models of cortical bone corresponding to the second set of simulations (Series II_RL) focusing on the regions of the occurrence of RL: (**a**) homogeneous bone, 1 RL with center coordinates (x, y) = (20, 2); (**b**) homogeneous bone, 1 RL with center coordinates (x, y) = (20, 3); (**c**) homogeneous bone, 1 RL with center coordinates (x, y) = (20, 1); (**d**) homogeneous bone, 1 RL with center coordinates (x, y) = (27, 2) mm; (**e**) homogeneous bone, 1 RL with center coordinates (x, y) = (20, 2) and double the diameter; (**f**) B_Po5_RL1; (**g**) B_Po5_RL3; (**h**) B_Po5_RL5.

3.3. Ultrasound Configuration

Simulations of ultrasound propagation were performed in the tangential direction (Figure 1a). Calculations were conducted by placing one point source and 14 point receivers: (a) at a distance 2 mm from the upper cortical layer (non-contact transducers); (b) directly onto the cortical cortex (implanted transducers). The distance between two successive receivers was 2.48 mm implying that the distance between the emitter and the receivers ranges from 1.984 to 34.224 mm (receiver 1 (R1) – receiver 14 (R14)). The width of each transducer was equal to 11 elements. The use of a linear array of transducers is in agreement with exsisting ultrasonic devices for bone characterization such as the bidirectional device in [24].

3.4. Determination of the Ultrasonic Wave Propagation Path and Velocity

The "local FAS velocities" correspond to small propagation paths and were calculated as the difference in the FAS arrival time of the signals from successive receivers (distance 2.48 mm). In this way, we intend to estimate the structural changes along small regions of cortical bone due to porosity variations or to the presence of larger pores. When multiple receivers are considered, the change in the FAS arrival time Δt (μs) can be calculated using $\Delta t(x) = t(x_i) - t(x_{i-1})$, where i is the number of the receiving position, $t(x_i)$ is the arrival time of the signal at receiver i and $t(x_{i-1})$ is the arrival time at the previous receiver [25]. For the detection of the FAS a threshold was applied to the receiving waveforms corresponding to the identification of the first signal extremum (Figure 2) [26].

3.5. Boundary Conditions

The cortical cortex was surrounded by soft tissues and perfectly matched layers (PML) were used to limit spurious reflections from the boundaries simulating an infinitely long model. The PML efficiency was set to 80 dB which means that the wave reflected by the PML is expected to be 80 dB below the amplitude of a incident wave, in the case of normal incidence [22,23].

3.6. Numerical Simulation and Signal Analysis in the Time Domain

The numerical solution of the 2D wave propagation problem was carried out by using an FDTD code [22,23]. The computations are based on a system of elastodynamic equations corresponding to the propagation of mechanical waves in continuous media which obey Hooke's law and are expressed as:

$$\rho(x)\frac{\partial u_i}{\partial t}(x,t)=\sum_{j=1}^{d}\frac{\partial T_{ij}}{\partial x_j}(x,t)+f_i(x,t) \tag{1}$$

$$\frac{\partial T_{ij}}{\partial t}(x,t)=\sum_{k=1}^{d}\sum_{l=1}^{d}c_{ijkl}(x)\frac{\partial u_k}{\partial x_l}(x,t)+\theta_{ij}(x,t) \tag{2}$$

where subscripts i={1,…,d} refer to the direction of space and d to the space dimension (d =2 for 2D computational models), x and t are the space and time variables, $\rho(x)$ is the mass density and c(x) is the fourth-order rigidity tensor. These parameters entirely define the material properties and geometry of the medium. u_i (x, t) are the vector components of the particle velocity field, T_{ij} (x, t) are the components of the stress tensor, while f_i denote the vector components of force sources and θ_{ij} denote the tensor components of strain rate sources.

A Hanning pulse was used as the excitation signal including four sinusoidal cycles [27]. The examined central angular frequencies were 0.5 and 1 MHz. The duration of the simulations was 25 μs. The accuracy of the solution depends on the relation between the element size and the wavelength. According to the software requirements, the stability condition is defined as:

$$\Delta t\leqslant\frac{1}{\sqrt{d}}\frac{\Delta x}{c_{max}} \tag{3}$$

where Δt and Δx denote the time and spatial steps used to approximate time or spatial derivatives, c_{max} is the largest speed of sound amongst all the simulated materials (c_{max} = c_{bone} = 3939 m/s), and d refers to the space dimension (d = 2). The grid size was set to 8 μm, and a convergence study is presented in the following subsection to demonstrate the accuracy of the findings.

3.6.1 Convergence Study

The grid size must be dense enough to ensure an accurate numerical solution and to minimize the computational cost. In addition, it must be small enough in comparison to the diameter of the pores. The convergence study is based on the numerical model B_Po5 which integrates the smallest scatterers among all the geometries of Table 2. Two transducers were considered which were placed directly onto the cortical surface and their distance was set to 20 mm. FDTD simulations were conducted and the FAS velocities were calculated for the central excitation frequency 1 MHz and the grid step sizes 4, 8, and 16 μm. The calculated FAS values were 3499 m/s for 4 μm, 3495 m/s for 8 μm, and 3489 m/s for 16 μm. It can be concluded that the 8-μm grid size is a good compromise for an accurate numerical solution and less time consuming simulations which is in agreement with [2].

3.7. Statistical Analysis

A statistical analysis was conducted and the results are presented as local mean FAS velocity ± standard error. Linear regression analysis and one-way analysis of variance (one-way ANOVA) were used to evaluate changes in porosity. The statistical findings were considered significant for *p*-values less than 0.05 [9].

4. RESULTS

Figure 4, Figure 5, Figure 6, Figure 7, Figure 8, Figure 9, Figure 10, Figure 11 and Figure 12 present the results derived from the two simulation sets of Figure 1 and Figure 3 and the central excitation frequencies 0.5 and 1 MHz. As multiple receivers were considered and small propagation paths were examined, the horizontal axis represents the distance of each receiving element from the starting point of the plate with coordinates $(x, y) = (0, 0)$. First, in Subsection 4.1, we present the results derived for porosity changes from 0% to 16%, and then, in Subsection 4.2, the findings corresponding to the occurrence of a single or a cluster of RLs in small propagation paths at different depths, lengths and sizes.

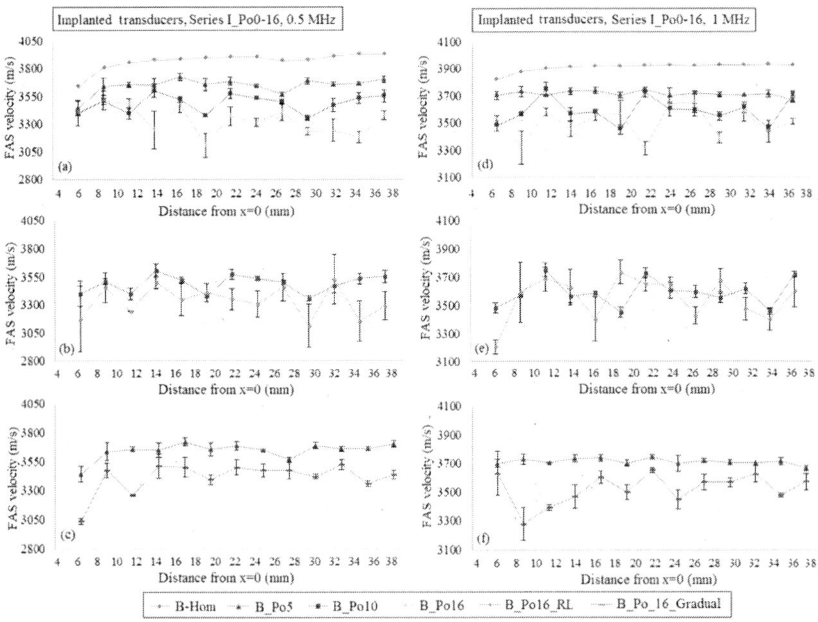

Figure 4. The potential of the mean local FAS velocity to detect changes in cortical porosity from 0%–16%. The diagrams correspond to the first set of simulations under the assumption of implanted transducers and the excitation frequencies: **(a)**–**(c)** 0.5

MHz; and **(d)**–**(f)** 1 MHz. The standard error bars demonstrate the FAS velocity variation for the three numerical models established for each porosity scenario.

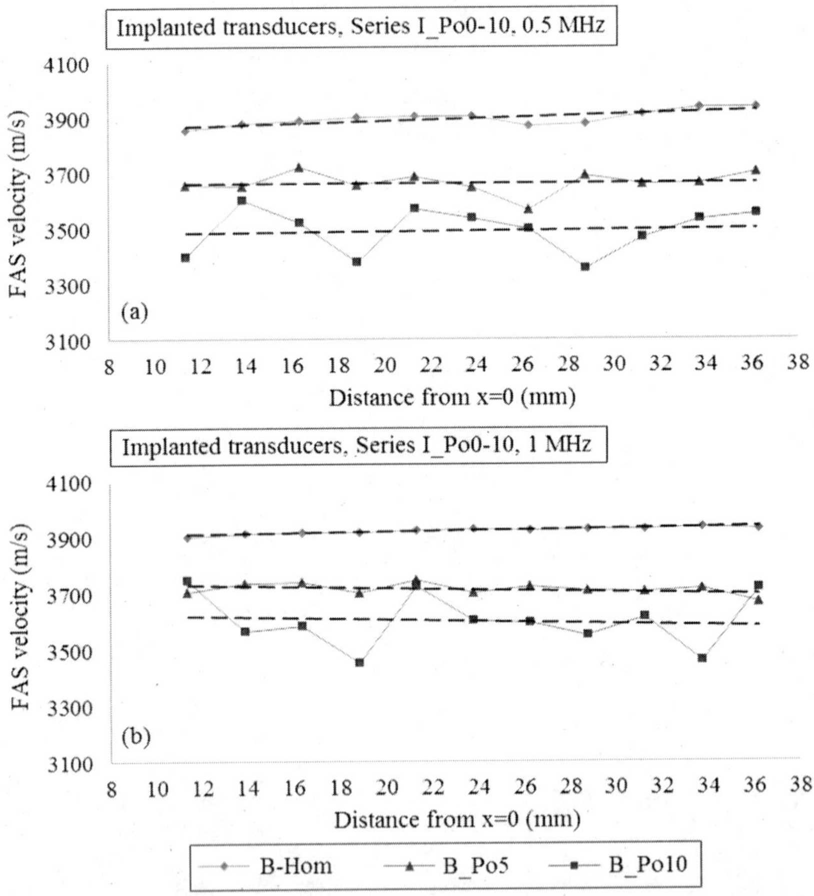

Figure 5. Linear fit derived from multiple receiving positions for cortical porosities from 0 to 10%. The diagrams correspond to the first set of simulations under the assumption of implanted transducers and the excitation frequencies: **(a)** 0.5 MHz; and **(b)** 1 MHz. The corresponding equations, coefficients of determination and root mean square errors are presented in Table 3 and Table 4.

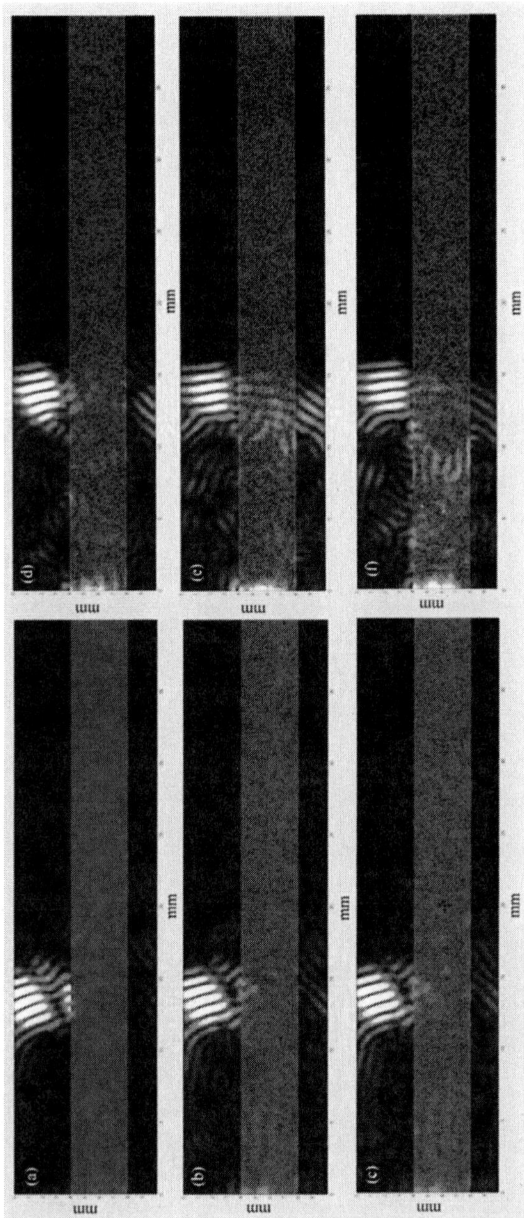

Figure 6. Snapshots of wave propagation for the first set of simulations using non-contact transducers for the excitation frequency of 1 MHz and time instant 10 μs for the cases: (**a**) B_Hom; (**b**) B_Po5; (**c**) B_Po5_RL5; (**d**) B_Po10; (**e**) B_Po16; (**f**) B_Po16_RL.

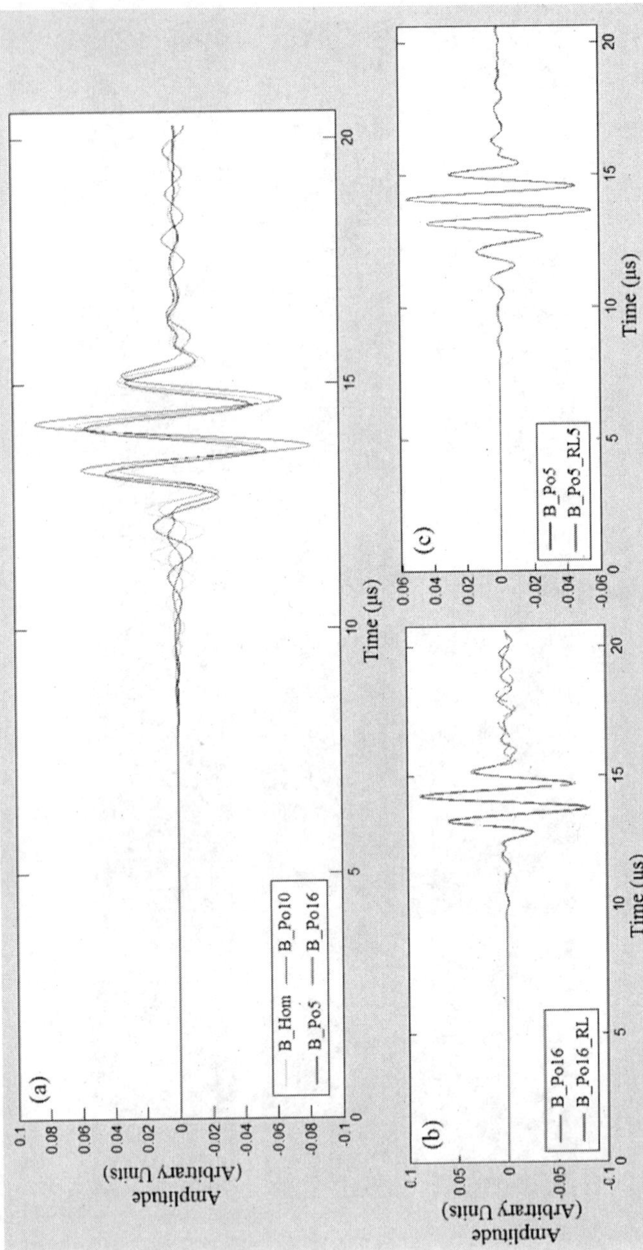

Figure 7. Waveforms derived from receiver R8 for the excitation frequency 1 MHz and simulation time 20 μscomparing the examined cases: (**a**)B_Hom, B_Po5, B_Po10 and B_Po16; (**b**) B_Po16 and B_Po16_RL; (**c**) B_Po5 and B_Po5_RL5.

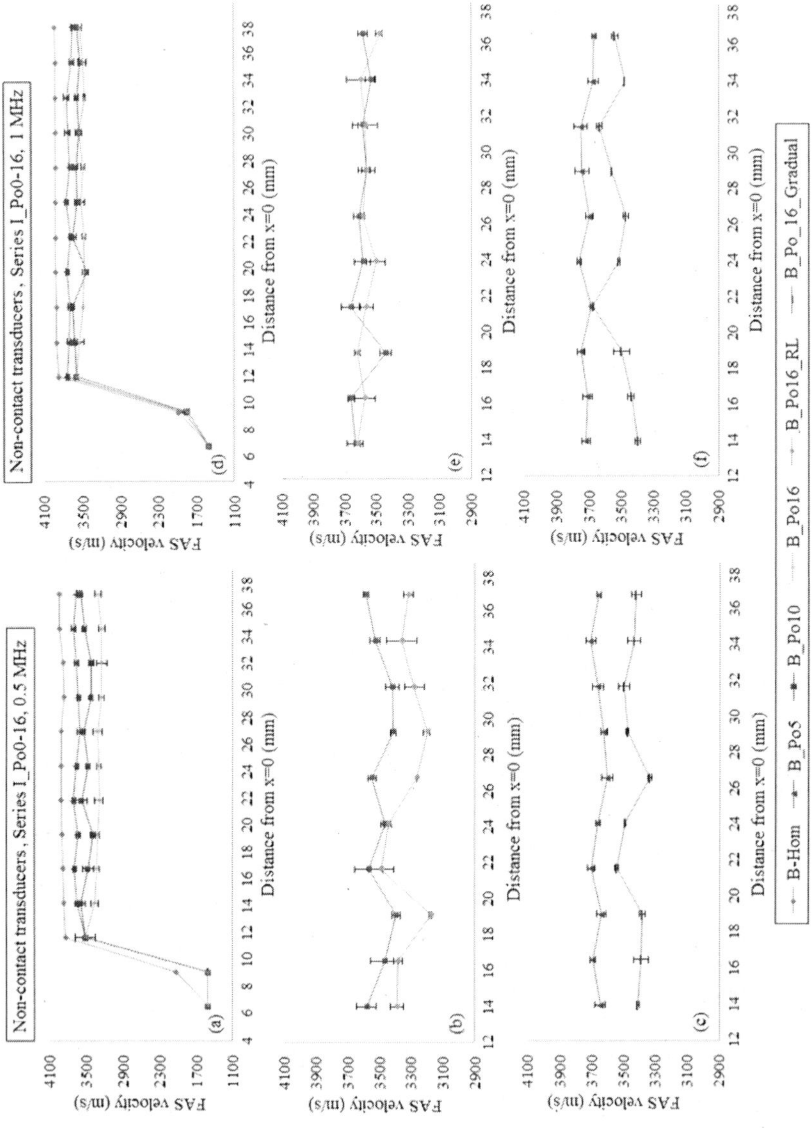

Figure 8. The potential of the mean local FAS velocity to detect changes in cortical porosity from 0% to 16%. The diagrams correspond to the first set of simulations when the transducers are placed at a distance of 2 mm from the cortical cortex. The results for the excitation frequencies: **(a)**–**(c)** 0.5 MHz; and **(d)**–**(f)** 1 MHz are presented. The standard error bars demonstrate the FAS velocity variation for the three numerical models established for each porosity scenario.

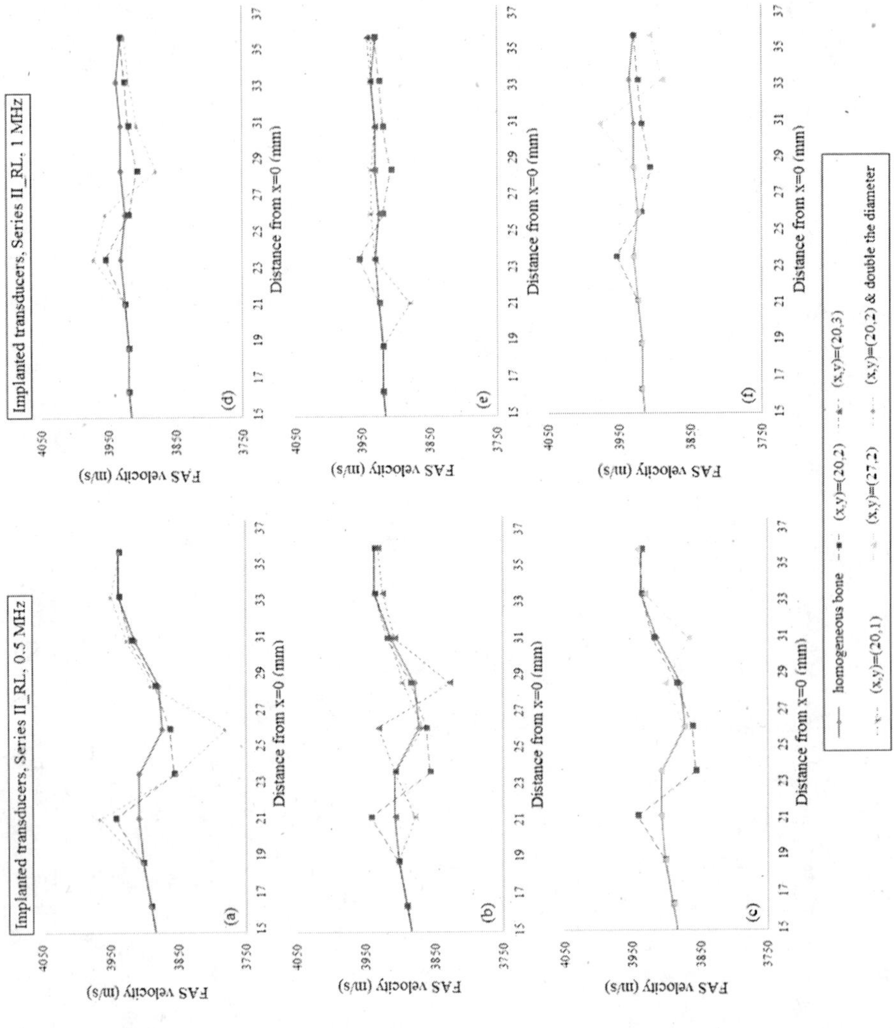

Figure 9. The potential of the local FAS velocity to detect the occurrence of a single RL considering cortical bone as a homogeneous medium (geometries illustrated in Figure 3a–e). The diagrams correspond to the second set of simulations under the assumption of implanted transducers and the excitation frequencies: **(a)**–**(c)** 0.5 MHz; and **(d)**–**(f)** 1 MHz. The legend describes the center coordinates of the RL.

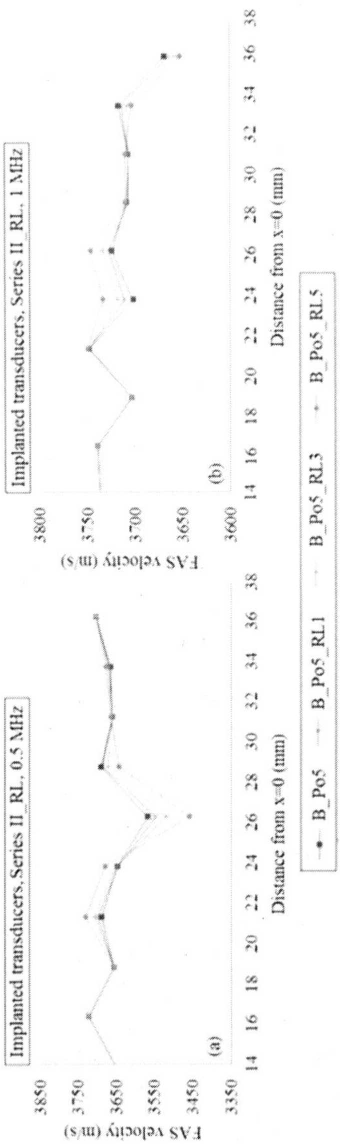

Figure 10. The potential of the local FAS velocity to detect the occurrence of a single or a cluster of RL considering cortical bone as a nonhomogeneous medium for the cases B_Po5, B_Po5_RL1, B_Po5_RL3 and B_Po5_RL5 (geometries illustrated in Figure 3f–h). The diagrams correspond to the second set of simulations under the assumption of implanted transducers and the excitation frequencies: (**a**) 0.5 MHz; and (**b**) 1 MHz.

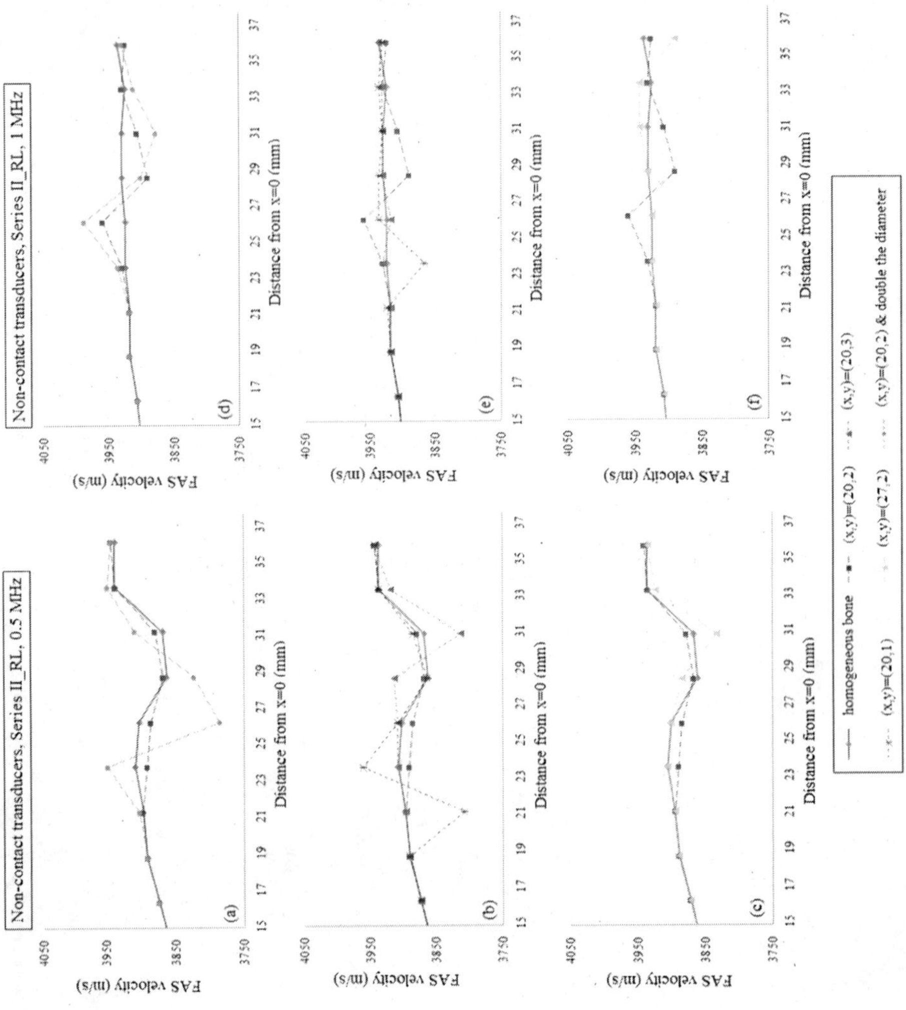

Figure 11. The potential of the local FAS velocity to detect the occurrence of a single RL considering cortical bone as a homogeneous medium (geometries illustrated in Figure 3a–e). The diagrams correspond to the second set of simulations when the transducers are placed at a distance of 2 mm from the cortical cortex. The results for the excitation frequencies : **(a)**–**(c)** 0.5 MHz; and **(d)**–**(f)** 1 MHz are presented. The legend describes the center coordinates of the RL.

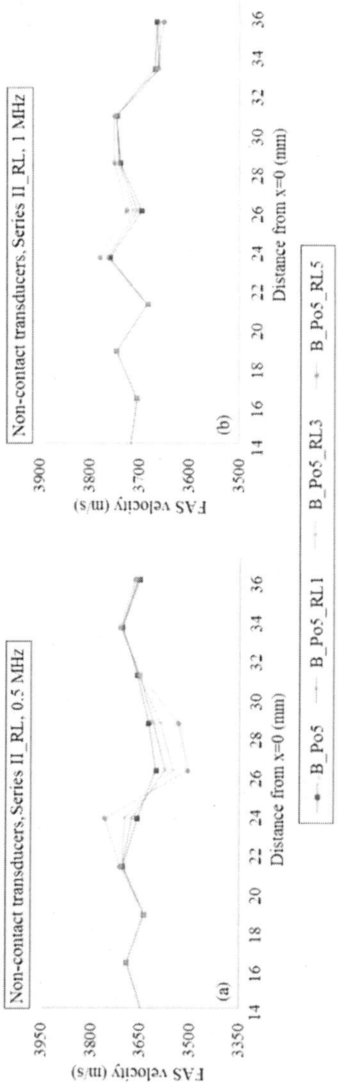

Figure 12. The potential of the local FAS velocity to detect the occurrence of a single or a cluster of RL considering cortical bone as a nonhomogeneous medium for the cases B_Po5, B_Po5_RL1, B_Po5_RL3 and B_Po5_RL5 (geometries illustrated in Figure 3f–h). The diagrams correspond to the second set of simulations when the transducers are placed at a distance of 2 mm from the cortical cortex. The excitation frequencies: (**a**) 0.5 MHz; and (**b**) 1 MHz are examined.

4.1. First Set of Simulations (Series I_Po0-16)

4.1.1. Implanted Transducers

Figure 4 presents the FAS velocity variation profile from multiple implanted receivers if we consider cortical bone as a nonhomogeneous medium with porosity from 0 to 16%. The local FAS velocities were calculated as the mean values from the same receiving position of three simulation maps with different random distributions of the pores keeping the same microstructure characteristics. The mean FAS velocities and the standard error bars for the excitation frequency 0.5 MHz are presented in Figure 4a–c and for 1 MHz in Figure 4d–f. It can be observed that a porosity increase leads to the decrease of the FAS velocity variation profile, and this is more apparent for the excitation frequency 0.5 MHz.

More specifically, if we consider the case B_Hom as the reference case, the relative percentage differences in Figure 4a for the cases B_Po5, B_Po10 and B_Po16 were calculated in the range 4.32%–7.87%, 6.58%–13.53% and 6.41%–20.34%, respectively. These findings can be directly compared with Figure 4d. For the excitation frequency 1 MHz, the relative percentage differences for the cases B_Po5, B_Po10 and B_Po16 were from 3.14% to 6.67%, 3.91%–12.10% and 7.35%–15.56%, respectively. Then, Figure 4b,c,e,f present more realistic scenarios of pathology including both normal pores and RLs for the case B_Po16_RL (Figure 4b for 0.5 MHz and Figure 4e for 1 MHz) and the gradual formation of osteoporosis (Figure 4c for 0.5 MHz and Figure 4f for 1 MHz). In Figure 4b, a porosity increase up to 16% including the occurrence of RLs (B_Po16_RL) leads to a decrease of the local FAS velocities and the relative percentage differences were from 9.23% to 19.71% for the excitation frequency 0.5 MHz. In addition, we observe that there are few positions (e.g. Figure 4b, $x = 30$–32 mm) which show a minor increase of the FAS velocity locally despite the porosity increase. Figure 4e presents the FAS velocities for

B_Po16_RL and the excitation frequency 1 MHz. It can be seen that as the frequency increases it is more difficult to discriminate a specific FAS velocity behavior among the cases B_Po10 and B_Po16_RL. Finally, Figure 4c,f show the potential of the local FAS velocity to detect the gradual formation of osteoporosis. The diagram B_Po5 is depicted in the same diagram with B_Po16_Gradual as when osteoporosis is gradually formed the upper cortical layer has the microstructure of a healthy bone, while the porosity and pores' size increase starts from the endosteal region. The relative percentage differences for the case B_Po16_Gradual keeping B_Hom as the reference were calculated from 8.70% to 16.63% for the excitation frequency 0.5 MHz (Figure 4c) and from 5.02% to 15.49% for 1 MHz (Figure 4f), respectively.

Then, one-way ANOVA was conducted and the *p*-value was calculated. It was found to be lower than 0.05 ($p \approx 0$), implying that there are significant differences between group means. Linear regression analysis was also performed, as in some receiving positions of Figure 4 the mean FAS velocity was found to increase locally despite the porosity increase.Figure 5 shows that linear regression analysis is not influenced by local phenomena, and the examined cases can be discriminated. The corresponding equations, coefficients of determination (R^2) and root mean square errors (RMSE) are included in Table 3 and Table 4. All the results are not presented graphically to reduce complexity, as similar findings toFigure 5 were derived.

4.1.2. Transducers at a Distance of Two Millimeters from the Cortical Cortex

Figure 6 presents snapshots of ultrasound wave propagation for indicative cases considering the excitation frequency 1 MHz and time instant 10 μs. It is shown that, as porosity increases, multiple scattering mechanisms evolve (Figure 6d,e,f). In addition, it can be observed that the application of PML limits spurious reflections from the boundaries, but the phenomenon is not totally diminished.

Then, in Figure 7 the received waveforms are illustrated derived from receiver R8, which is placed above the region of the cluster of 5 RLs (Figure 3h) for indicative cases and the excitation frequency 1 MHz, It can be seen in Figure 7a that the increase of porosity induces a delay in the propagation of the signal if we compare the zero-crossing times. An increase in amplitude was also found with increasing the porosity. A comparison of the waveforms of the cases B_Po16 and B_Po16_RL is presented in Figure 7b, showing that the occurrence of the larger pores has an effect on the received waveforms. However, in Figure 7c, the waveforms for the cases B_Po5 and B_Po5_RL5 almost coincide, while no difference can be observed at the corresponding snapshots of Figure 6.

Figure 8 presents the mean FAS velocities and standard error bars when the transducers are placed at a distance of 2 mm from the cortical surface. The porosities from 0% to 16% are examined as in Figure 4. In Figure 8a,d, low FAS velocities are observed for the receivers which are placed before x = 12 mm due to the direct propagation of the FAS via the soft tissues.

More specifically, if we consider the case B_Hom as the reference case, the relative percentage differences in Figure 8a for 0.5 MHz and B_Po5, B_Po10 and B_Po16 were calculated in the range 4.71%–7.65%, 7.22%–12.83% and 12.90%–17.60%,respectively. For the excitation frequency 1 MHz (Figure 8d), the relative percentage differences for B_Po5, B_Po10 and B_Po16 were from 4.24% to 6.89%, 6.37%–12.19% and 8.78%–11.82%, respectively. Then, Figure 8b,e corresponds to the case B_Po16_RL and are directly compared with the case B_Po10. Keeping B_Hom as the reference case the relative percentage differences for B_Po16_RL were from 10.51 to 18.33% for 0.5 MHz (Figure 8b) and 7.17%–

11.62% for 1 MHz (Figure 8e). Thus, even if according to Figure 7b, the two waveforms seem to coincide for propagation time less then 10 μs, however the calculation of the FAS in small propagation paths shows quantitative differences when large pores are considered. The results for the gradual distribution of the pores are shown in Figure 8c,f. The relative percentage differences for B_Po16_Gradual were derived from 9.02% to 14.55% for the excitation frequency 0.5 MHz (Figure 8c) and from 6.31% to 12.95% for 1 MHz (Figure 8f), respectively. It should be mentioned that in Figure 8b,c,e,f, we do not present the whole propagation path, as the initial low values of Figure 8a,b corresponding to the direct FAS propagation through the water were neglected. In addition, linear regression analysis was performed and the results are depicted in Table 3 and Table 4. It was shown that the examined cases can be discriminated even for the diagrams of Figure 8e in which the mean local FAS velocities do not show a specific behavior with increasing the porosity.

Concerning one-way ANOVA, the *p*-value was lower than 0.05 ($p \approx 0$) ignoring the receiving positions near the source ($d < 12$ mm). The reason for not considering the first receivers in the calculations is that as it can be seen in Figure 8a,d for $d < 12$ mm, the FAS wave propagates directly in the soft tissues and low velocities are derived in comparison to the bulk longitudinal velocity of bone. This assumption was also considered for the cases of Figure 4 to keep the same conditions.

4.2. Second Set of Simulations (Series II_RL)

4.2.1. Implanted Transducers

This section presents the results corresponding to the occurrence of: (a) a single RL at different depths, lengths and sizes under the assumption of cortical homogeneous bone (Figure 9 and Figure

11); (b) a single or a cluster of RL taking into account cortical microstructure (Figure 10 and Figure 12). In Figure 9 and Figure 10, implanted transducers are used and the excitation frequencies 0.5 and 1 MHz are examined. Figure 9a–c depict the local FAS velocity variation profile for the excitation frequency 0.5 MHz and the presence of a single RL. For comparison purposes, all the diagrams include the calculated values for the intact bone (blue line) and the case of a single RL with center coordinates $(x, y) = (20, 2)$ (red line). All the diagrams start from a distance of 15 mm in the x axis as for lower cortical lengths all the curves coincide. In Figure 9a, we observe that for 0.5 MHz the occurrence of one RL in the center of the plate can be captured by the receivers which are placed directly above the region of interest showing an increase of the FAS velocity of 35 m/s compared to the homogeneous bone diagram. In addition, if we keep the same RL position and double the diameter of the pore (orange line), the receivers which are placed in the position x = 19–21 mm are the first to detect an increase of the FAS velocity of 58 m/s in comparison to the homogeneous bone value. In Figure 9b, we examine how the depth of the RL occurrence influences the FAS velocity. It can be seen that, if we place the RL at a higher cortical surface (grey line), the receivers which first capture the presence of a pore are placed again in the propagation path between x = 19–21 mm showing a velocity decrease of 29 m/s comparing to the reference case of homogeneous bone. On the other hand, if we increase the depth from the cortical surface (purple line), there is a delay in the identification of the RL by the receivers above the region of x = 25–21 mm showing an increase of the FAS velocity of 57 m/s. Finally, Figure 9c shows that if we change the RL position along the x axis, the receivers which are placed above the region of interest are the first to detect an increase of the FAS velocity of 23 m/s revealing the position of the RL formation.

Table 3. Linear fit of the mean first arriving signal (FAS) velocities derived from all the receiving positions and the excitation frequency 0.5 MHz.

	Implanted Transducers			Transducers at a Distance of 2 mm from the Cortical Surface		
	$y = ax + b$	R^2	RMSE	$y = ax + b$	R^2	RMSE
B_Hom	$y = 2.27x + 3844$	0.54	16.56	$y = 3.00x + 3815$	0.52	22.79
B_Po5	$y = 0.17x + 3661$	0.01	37.94	$y = 2.81x + 3577$	0.17	48.84
B_Po10	$y = 0.72x + 3475$	0.01	78.51	$y = 0.28x + 3489$	0.01	66.92
B_Po16	$y = -4.37x + 3417$	0.09	110.08	$y = -3.64x + 3385$	0.53	24.58
B_Po16_RL	$y = -4.09x + 3434$	0.06	123.45	$y = -3.23x + 3410$	0.06	92.93
B_Po16_Gradual	$y = 1.38x + 3409$	0.02	75.88	$y = 1.57x + 3397$	0.03	59.72

Table 4. Linear fit of the mean FAS velocities derived from all the receiving positions and the excitation frequency 1 MHz.

	Implanted Transducers			Transducers at a Distance of 2 mm from the Cortical Surface		
	$y = ax + b$	R^2	RMSE	$y = ax + b$	R^2	RMSE
B_Hom	$y = 1.08x + 3898$	0.80	4.26	$y = 1.65x + 3879$	0.80	6.37
B_Po5	$y = -1.41x + 3748$	0.28	17.82	$y = -1.81x + 3759$	0.19	29.16
B_Po10	$y = -1.54x + 3639$	0.02	93.23	$y = -1.78x + 3628$	0.06	57.31
B_Po16	$y = -1.36x + 3549$	0.01	100.58	$y = -0.06x + 3510$	0.01	30.84
B_Po16_RL	$y = -5.38x + 3702$	0.13	111.10	$y = -3.21x + 3647$	0.28	36.73
B_Po16_Gradual	$y = 3.96x + 3440$	0.15	72.96	$y = 4.49x + 3406$	0.16	73.61

Figure 9d–f presents the local FAS velocity variation profile for the excitation frequency 1 MHz and the presence of a single RL. In Figure 9d, it can be seen that the occurrence of a single RL in the center of the plate (red line) can be identified with a delay by the receivers which follow the region of interest (around $x = 23$ mm) showing an increase of the FAS velocity of 23 m/s. In addition, it was found that if we double the diameter of the RL (orange line), the same receivers are the first to detect an increase of the FAS velocity of 41 m/s. Then, Figure 9e indicates that by increasing the depth from the cortical surface (purple line), the larger pore cannot be detected as this diagram coincides with the case of homogeneous bone (blue line). On the other hand, the occurrence of the RL near the upper cortical surface (grey line) leads to a decrease of the FAS velocity of 46 m/s derived from the receivers directly above the region of interest. Finally, Figure 9f shows that if we change the position of the RL along the x axis, the receivers which are placed approximately at a distance of 3 mm after the pore are the first to capture an increase of the FAS velocity of 47 m/s.

Then, in Figure 10, we investigate the potential of the local FAS velocity to detect the occurrence of a single or a cluster of three or five large pores in the center of the plate taking into account the nonhomogeneous nature of cortical bone. Figure 10a presents the FAS velocity variation profile for the excitation frequency of 0.5 MHz, while Figure 10b the FAS values for 1 MHz considering the diagram of B_Po5 (red line) as the reference case. For the excitation frequency of 0.5 MHz the relative percentage differences were calculated in the range: (a) 0.04%–0.28% for B_Po5_RL1, (b) 0.04%–1.36% for B_Po5_RL3 and (c) 0.04%–3.08% for B_Po5_RL5. It can be seen that the first receivers which can identify the RL occurrence are placed above the region of interest ($x = 20$–22 mm). A 2-mm delay in the detection of the larger pores is observed in Figure 10b for 1 MHz as they are first identified from the receivers in the region $x = 22$–24 mm. Lower relative percentage differences were calculated compared to 0.5 MHz from: (a) 0.05%–0.24% for B_Po5_RL1, (b), 0.23%–0.42% for B_Po5_RL3, and (c) 0.05%–0.85% for B_Po5_RL5.

4.2.2. Transducers at a Distance of Two Millimeters from the Cortical Cortex

Figure 11a–c present the local FAS velocity variation profile for the excitation frequency 0.5 MHz and the presence of a single RL by placing the transducers at a distance of 2 mm from the cortical cortex. All the diagrams include the case of intact homogeneous bone (blue line) and the case of a single RL with center coordinates $(x, y) = (20, 2)$ (red line). In Figure 11a, it can be seen that, for 0.5 MHz, the occurrence of one RL in the center of the plate can be captured by the receivers which are placed directly after the region of interest (between x = 21–23 mm) showing a decrease of the FAS velocity of 17 m/s compared to the homogeneous bone diagram. Then, if we keep the same RL position and double the diameter of the RL (orange line) the same receivers are the first to detect an increase of the FAS velocity of 41 m/s in comparison to the homogeneous bone value. Figure 11b shows the impact of the RL depth on the FAS values. We observe that if we place the RL at a higher cortical layer (grey line), the receivers which first capture the presence of a pore are placed directly above the region of interest showing a velocity decrease of 89 m/s comparing to the reference case of homogeneous bone. However, if we increase the depth from the cortical cortex (purple line), the receivers which are placed at a distance approximately 5 mm from the RL are the first to identify its occurrence showing an increase of the FAS velocity of 51 m/s. Finally, Figure 11c shows that if we change the RL position along the x-axis, the receivers which are placed above the region of interest are the first to capture an increase of the FAS velocity of 23 m/s revealing the position of the larger pore.

Additionally, Figure 11d–f present the local FAS velocity variation profile for the excitation frequency 1 MHz and the presence of a single RL. Specifically, Figure 11d shows that the occurrence of a single RL in the center of the plate (red line) can be identified with a delay by the receivers which follow the region of interest (around $x = 23–25$ mm) showing an increase of the FAS velocity of 35 m/s. In addition, it can be observed that if we double the diameter of the RL (orange line) the same receivers are the first to identify an increase of the FAS velocity of 65 m/s. In Figure 11e, we observe that if we increase the depth from the cortical surface (purple line) the larger pore cannot be detected as this diagram almost coincides with the case of homogeneous bone. On the contrary, the occurrence of the RL near the upper cortical surface (grey line) leads to a decrease of the FAS velocity of 57 m/s, and it can be identified with a small delay by the receivers placed above the propagation path between $x = 21–23$ mm. Finally, Figure 11f shows that, if we change the position of the RL along the x-axis, the receivers which are placed between $x = 29–31$ mm are the first to detect an increase of the FAS velocity of 17 m/s.

Then, in Figure 12, we investigate the potential of the local FAS velocity variation profile to identify the occurrence of a single or a cluster of three or five large pores in the center of the plate taking into account the cortical microstructure. Figure 12a illustrates the local FAS velocities for the excitation frequency 0.5 MHz, while Figure 12b shows the FAS values for 1 MHz considering B_Po5 (red line) as the reference case. For the excitation frequency 0.5 MHz, the relative percentage differences were calculated in the range: (a) 0.05%–0.75% for B_Po5_RL1, (b) 0.05%–1.41% for B_Po5_RL3 and (c) 0.18%–2.71% for B_Po5_RL5. It can be observed that the first receivers which capture the RL are placed between $x = 21–23$ mm showing a short delay. A 4-mm delay in the detection of the larger pores can be seen in Figure 10b for 1 MHz and lower relative percentage differences were calculated compared to 0.5 MHz from: (a) 0.04–0.32% for B_Po5_RL1, (b)

0.14–0.46% for B_Po5_RL3, (c) 0.05%–0.79% for B_Po5_RL5, respectively.

5. DISCUSSION

This work presents a 2D parametric and systematic computational study aiming to investigate the effect of cortical porosity on ultrasound wave propagation in healthy and osteoporotic long bones. In comparison to previous studies [2,14,26], we established more realistic scenarios, as this is the first time that the distribution of the pores was randomized including normal pores, as well as pores with diameters larger than the Haversian canal size. Moreover, the analysis was not limited to a small segment of cortical bone with the presence of only one large basic multicellular unit as in [14]. The gradual formation of osteoporosis was also considered by simulating more physiological distributions of pores with larger pores in the endosteal cortex and smaller ones in the periosteum. The excitation frequencies 0.5 MHz and 1 MHz were examined to investigate which frequency is more sensitive to detect changes in cortical porosity and the occurrence of RLs. Numerical simulations of wave propagation were performed in the tangential direction and various microstructural models mimicking normal and pathological tissue states were investigated. In comparison to traditional FAS velocity measurements using one source and one receiver, we performed calculations of "local" FAS values in small successive propagation paths along the cortical cortex in order to establish possible guidelines for the interaction of ultrasound with cortical microstructure.

Initially, in Figure 4a,d, and Figure 8a,d, we examined the FAS velocity variation profile along cortical bone for microstructural models with porosities from 0 to 16%. It was made clear that as the porosity increases, the FAS velocities calculated in small propagation paths decrease for both the examined frequencies. However, the relative percentage differences calculated considering B_Hom as the reference case were higher for 0.5 MHz comparing to 1 MHz. Moreover,Figure 7a shows a delay in the time of arrival with increasing the porosity followed by an amplitude increase. This may be explained by the reflection mechanisms, which enhance the received signal for higher porosities. However, despite the significance of the calculation of the FAS velocities from the first signal extremum, the evaluation of the whole waveform would be a possible direction for future research based on guided wave analysis or on the waveform rectification followed by numerical integration via the trapezoidal method (presented in Section 2).

Additionally, the use of implanted transducers led to higher relative percentage differences compared to the use of non-contact transducers, while a delay in the detection of a single or a cluster of RL was observed for non-contact transducers. This can be attributed to the attenuation mechanisms induced by the presence of the soft tissues in the propagation path. The attenuation mechanisms include the evolution of absorption and scattering phenomena. Scattering is induced in bone due to the presence of the pores while soft tissues are major sources of acoustic wave absorption. Despite the accuracy of the FDTD method, which was demonstrated via a convergence study, a limitation of the present work is the simulation weakness of the software to account for attenuation effects resulting from the viscoelasticity [2]. In addition, it should be mentioned that the use of implanted transducers is considered as an invasive procedure for osteoporosis and was examined only for comparison purposes with the non-contact transducers. In clinical practice, their insertion is easy for fracture healing assessment [12,27], but, for the evaluation of osteoporosis, this configuration is not applied during *in vivo*measurements.

Concerning the case of pathology with porosity of 16%, two additional scenarios were investigated to account for the presence of pores with different diameters as well as for the gradual formation of osteoporosis. It was found that the FAS velocity variation profile shows a decrease along the cortical cortex for both B_Po16_RL and B_Po16_Gradual for 0.5 MHz. On the other hand, the excitation frequency of 1 MHz was not convenient for B_Po16_RL (Figure 4e and Figure 8e) as no specific FAS velocity tendency was observed, while, for B_Po16_Gradual, it was sensitive to the gradual porosity increase. A direct comparison of the received signals from B_Po16 and B_Po16_RL was also included in Figure 7b showing a difference in the received signals due to the occurence of large pores.

Additionally, a statistical analysis was conducted based on the estimation of the standard errors, one-way ANOVA and linear regression analysis. It should be mentioned that the receiving positions near the source ($d < 12$ mm) were ignored as in Figure 8a,d, we observe that for short distances between the transducers, low velocities are calculated, revealing that the FAS propagates directly via water. The calculated *p*-values were lower than 0.05 indicating that there are statistically significant differences between group means as the porosity increases from 0% to 16%. The low *p*-values can be explained by the high dispersion of the FAS velocity derived from serial measurements for porosity of 16%. Linear regressions analysis revealed a specific FAS velocity trend even for the cases of Figure 4e and Figure 8e. Nevertheless, low R^2 values were derived implying that this analysis could be used only for supplementary observations following traditional local and successive FAS velocity measurements. The establishment of three numerical models for each case is another parameter which requires further discussion. The use of three models is considered as a good compromise to reduce the computational cost and derive accurate conclusions taking into account the calculated standard error values. More specifically, for all the examined cases after three runs, the maximum FAS velocity variation was far less than 5%, thus we decided not to make more runs to keep calculations and simulations simpler.

Then, the case of normal remodeling was examined including the presence of a single or a cluster of RL. Specifically, it was shown that for implanted transducers and 0.5 MHz, as well as for both the hypothetic case of homogeneous and the realistic of nonhomogeneous bone, the presence of one or more RLs can be identified by the receivers which are placed directly above the region of interest revealing the effectiveness of the FAS wave as an indicator. The same observation was made for all the sizes, numbers, depths and lengths of the RL. However, the use of 1 MHz or the transducers' placement into the water led to the detection of larger pores with a short delay when the RLs are placed in the center of the plate (approximately 4 mm) or near the upper cortical cortex (approximately 2 mm). It should be emphasized that the occurrence of a single RL near the lower surface of cortical bone could not be identified when the excitation frequency 1 MHz was applied irrespective of the transducers' configuration. In addition, it was observed that as the receivers are removed from the region of interest the calculated values gradually approximate the values of healthy bone. Nonetheless, the combination of low and high frequencies for the detection of larger pores could be useful, as low frequencies can detect the region of interest along the x axis and higher frequencies can reveal the depth of the pores along the y-axis. For example, in Figure 9b,e, when the RL is placed near the upper cortical layer (grey line), for both frequencies, we can detect the large pore in the same position implying that the pore is near the cortical upper surface. If we place the RL in the center of the plate (red line), the use of 0.5 MHz frequency leads to the detection of the exact x position of the RL, but 1 MHz shows a delay. This reveals that the pore is placed deeper. Finally, Figure 9e shows that for 1 MHz we cannot detect the RL at y = 3 (purple line), indicating in combination with 0.5

MHz findings that a large pore exist, but this pore is positioned near the lower surface of the cortical cortex. Thus, even from the limitations of 1 MHz, we may derive specific information concerning the depth of the RL interpreted in combination with 0.5 MHz findings when the occurrence of large basic multicellular units is a factor of major concern.

A possible explanation for the low sensitivity of 1 MHz is that, when the wavelengths are comparable to or smaller ($\lambda_{0.5MHz}$ = 7.88 mm, λ_{1MHz} = 3.94 mm), the thickness of the cortical cortex (4 mm), the FAS wave propagates as a subsurface, lateral wave which travels at the bulk longitudinal velocity of the medium and cannot reflect changes occurring at deeper cortical layers. The calculation of the wavelenths depends on the frequency and the bone's bulk velocity which is 3939 m/s considering the material properties of Table 1. However, bone velocity may vary from 3500 m/s to 4100 m/s depending on the type and anatomic region of bone as well as cortical thickness. According to the literature [15,28], the use of low excitation frequencies (0.2 MHz) has been chosen to yield information on the correlation of the FAS velocity with changes in cortical thickness due to osteoporosis, while it was reported in [28] that the FAS velocities for high frequencies such as 1 MHz were weakly correlated with cortical thicknesses. Therefore, future research should address the question of which frequency range below 0.5 MHz is more convenient for the assessment of osteoporosis considering the occurrence of pores with different sizes.

Concerning the case of homogeneous bone, it was observed that, for 1 MHz, the FAS velocity along the propagation path increases slightly, approximating the bone's bulk velocity (Figure 9d–f), while this is not the case for 0.5 MHz in Figure 9a–c. The observed behavior at 0.5 MHz can be attributed to the fact that the wavelength is almost twice the cortical thickness implying that more complex wave propagation phenomena evolve and the investigation of the propagation of guided waves could convey significant information. On the other hand, for 1 MHz, the FAS propagates as a lateral wave.

Finally, several assumptions were made which require further numerical research and improvements. First of all, although two dimensional geometries were developed, they incorporated various porosity scenarios. Moreover, changes in cortical thickness in cases of pathology in combination with changes in cortical porosity were not considered. Imaging modalities could be also exploited to establish more realistic computational models [29,30]. In addition, the use of point transducers should be mentioned as in experimental procedures the transducers have a finite size. It was reported in [4] that the choice of using point elements instead of finite-sized ones did not affect significantly the recorded ultrasound velocities. According to our previous study using numerical models from SAM images [31], the application of transducers with a finite size larger than the RL diameter could provide significant information concerning the occurrence of a region of RLs. However, further numerical research is needed to investigate the impact of the transducers' size in relation to the scatterers' size on the results. Furthermore, according to [32], the majority of patients with osteoporosis have changes in the mineralization pattern. In particular, reduction of average mineral content was described when compared with agematched controls or when compared with reference values for healthy adults. Quantitative ultrasound was used in [33] for bone characterization and the role of bone mineral density as a major determinant of acoustic properties was highlighted as well as the significance of density-independent relationships with bone microarchitecture. However, according to [34], the main cause of sound attenuation coefficient in bone is porosity, while matrix stiffness has a minor effect. In addition, it was reported in [19] that in aged women the changes in porosity prevail over those of of matrix elasticity to drive the variations of the bone mesoscopic elasticity.

Nevertheless, the results clearly indicated that the excitation frequency 0.5 MHz is more sensitive to detect changes in cortical porosity and the occurrence of RLs. In addition, it was shown that the calculation of the FAS velocity in small propagation paths could potentially provide significant quantitative information for the early diagnosis of osteoporosis and detection of pathologic regions with larger pores with a high risk for fracture.

6. CONCLUSIONS

In this work, we presented a systematic and parametric computational study on ultrasound wave propagation in healthy and osteoporotic bones. The FAS velocity was derived in small propagation paths to extract quantitative criteria for the diagnosis and assessment of osteoporosis taking into consideration not only total changes in cortical porosity, but also the occurrence of larger pores and various microstructural features. The role of the excitation central frequency and the impact of the transducers' positions on the FAS velocity variation profile were examined. For the first time, the gradual formation of osteoporosis was also considered. It was shown that the local FAS velocities derived from a proper combination of low and high frequencies may contribute to the prevention or monitoring of osteoporosis and reveal potential pathologic regions with larger pores at an early stage of the disease. Further numerical research is suggested on challenges for clinical application and more realistic scenarios using scanning acoustic microscopy images.

ACKNOWLEDGMENTS

This work is part funded by the State Scholarships Foundation and the German Academic Exchange Service and by Deutsche Forschungsgemeinschaft (grant Ra1380/9-1).

AUTHOR CONTRIBUTIONS

Vassiliki T. Potsika worked on all the sections of the paper. Also, she was mainly responsible for the writing of the manuscript. Konstantinos N. Grivas and Theodoros Gortsas contributed to the development of the numerical models in Matlab and provided the algorithms for the random distribution of the scatterers. Gianluca Iori mainly contributed to Section 2 concerning the benchmark problem using both k-Wave and Simsonic. Vasilios C. Protopappas and Kay Raum set the parameters for the development of innovative numerical models based on realistic scenarios. They provided basic guidelines concerning: (a) the final porosities which were examined describing different stages of the pathology and (b) the selection of the proper numerical tools. Also, they contributed to the presentation of the results and structure of the paper. Demosthenes Polyzos and Dimitrios I. Fotiadis contributed to the assessment of the final results and the evaluation of: (a) the progress of the students and their individual as well as joint projects (b) the writing of the manuscript.

CONFLICTS OF INTEREST

The authors wish to declare that there is no conflict of interest.

REFERENCES

1. Laugier, P.; Haïat, G. *Bone Quantitative Ultrasound*; Springer: New York, NY, USA, 2010.

2. Rohde, K.; Rohrbach, D.; Gluer, C.C.; Laugier, P.; Grimal, Q.; Raum, K.; Barkmann, R. Influence of Porosity, Pore Size, and Cortical Thickness on the Propagation of Ultrasonic Waves Guided Through the Femoral Neck Cortex: A Simulation Study. *IEEE Trans. Ultras. Ferroel. Freq. Control* **2014**, *61*, 302–313.

3. Grondin, J.; Grimal, Q.; Engelke, K.; Laugier, P. Potential of first arriving signal to assess cortical bone geometry at the hip with QUS: A model based study. *Ultrasound Med. Biol.* **2010**, *36*, 656–666.

4. Moilanen, P.; Talmant, M.; Bousson, V.; Nicholson, P.H.F.; Cheng, S.; Timonen, J.; Laugier, P. Ultrasonically determined thickness of long cortical bones: two-dimensional simulations of *in vitro* experiments. *J. Acoust. Soc. Am.***2007**, *122*, 1818–1826.

5. Malo, M.K.H.; Rohrbach, D.; Isaksson, H.; Töyräs, J.; Jurvelin, J.S.; Tamminena, I.S.; Kröger, H.; Raum, K. Longitudinal elastic properties and porosity of cortical bone tissue vary with age in human proximal femur. *Bone* **2013**, *53*, 451–458.

6. Xu, K.; Liu, D.; Ta, D.; Hu, B.; Wang, W. Quantification of guided mode propagation in fractured long bones.*Ultrasonics* **2014**, *54*, 1210–1218.

7. Papacharalampopoulos, A.; Vavva, M.; Protopappas, V.; Polyzos, D.; Fotiadis, D.I. A numerical study on the propagation of Rayleigh and guided waves in cortical bone according to Mindlin's Form II gradient elastic theory. *J. Acoust. Soc. Am.* **2011**, *130*, 1060–1070.

8. Machado, C.B.; Pereira, W.C.D.A.; Granke, M.; Talmant, M.; Padilla, F.; Laugier, P. Experimental and simulation results on the effect of cortical bone mineralization in ultrasound axial transmission measurements: A model for fracture healing ultrasound monitoring. *Bone* **2011**, *48*, 1202–1209.

9. Rohrbach, D.; Lakshmanan, S.; Peyrin, F.; Langer, M.; Gerisch, A.; Grimal, Q.; Laugier, P.; Raum, K. Spatial distribution of tissue level properties in human a femoral cortical bone. *J. Biomech.* **2012**, *45*, 2264–2270.

10. Potsika, V.T.; Vavva, M.G.; Protopappas, V.C. Computational Modeling of Ultrasound Wave Propagation in Bone. In*Computational Medicine in Data Mining and Modeling*; Springer: New York, NY, USA, 2014; pp. 349–376.

11. Moreau, L.; Minonzio, J.G.; Talmant, M.; Laugier, P. Measuring the wavenumber of guided modes in waveguides with linearly varying thickness. *J. Acoust. Soc. Am.* **2014**, *135*, 2614–2624.

12. Protopappas, V.C.; Fotiadis, D.I.; Malizos, K.N. Guided ultrasound wave propagation in intact and healing long bones. *Ultras. Med. Biol.* **2006**, *32*, 693–708.

13. Malo, M.K.H.; Töyräs, J.; Karjalainena, J.P.; Isakssonc, H.; Riekkinena, O.; Jurvelin, J.S. Ultrasound backscatter measurements of intact human proximal femurs—Relationships of ultrasound parameters with tissue structure and mineral density. *Bone* **2014**, *64*, 240–245.

14. Bourgnon, A.; Sitzer, A.; Chabraborty, A.; Rohde, K.; Varga, P.; Wendlandt, R.; Raum, K. Impact of microscale properties measured by 50-MHz acoustic microscopy on mesoscale elastic and ultimate mechanical cortical bone properties. In Proceedings of IEEE International Ultrasonics Symposium, Chicago, IL, USA, 3–6 September 2014; pp. 636–638.

15. Nicholson, P.; Moilanen, P.; Kärkkäinen, T.; Timonen, J.; Cheng, S. Guided ultrasonic waves in long bones: modelling, experiment and application. *Physiol. Meas.* **2002**, *23*, 755–768.

16. Bossy, E.; Talmant, M.; Laugier, P. Effect of bone cortical thickness on velocity measurements using ultrasonic axial transmission: a 2d simulation study. *J. Acoust. Soc. Am.* **2002**, *112*, 297–307.

17. Moilanen, P.; Talmant, M.; Nicholson, P.H.F.; Cheng, S.; Timonen, J.; Laugier, P. Ultrasonically determined thickness of long cortical bones: Three-dimensional simulations of *in vitro* experiments. *J. Acoust. Soc. Amer.* **2007**, *122*, 2439–2445.

18. Grimal, Q.; Rohrbach, D.; Grondin, J.; Barkmann, R.; Gluer, C.C.; Raum, K.; Laugier, P. Modeling of femoral neck cortical bone for the numerical simulation of ultrasound propagation. *Ultrasound Med. Biol.* **2013**, *40*, 1–12.

19. Granke, M.; Grimal, Q.; Saïed, A.; Nauleau, P.; Peyrin, F.; Laugier, P. Change in porosity is the major determinant of the variation of cortical bone elasticity at the millimeter scale in aged women. *Bone* **2011**, *49*, 1020–1026.

20. Thompson, D.D. Age changes in bone mineralization, cortical thickness, and haversian canal area. *Calcif. Tissue Int.* **1980**, *31*, 5–11.

21. Treeby, B.E.; Jaros, J.; Rendell, A.P.; Cox, B.T. Modeling nonlinear ultrasound propagation in heterogeneous media with power law absorption using a k-space pseudospectral method. *J. Acoust. Soc. Am.* **2012**, *131*, 4324–4336.

22. Bossy, E.; Grimal, Q. Numerical methods for ultrasonic bone characterization. In *Bone Quantitative Ultrasound*; Laugier, P., Haiat, G., Eds.; Springer: New York, NY, USA, 2010; pp. 181–228.

23. Bossy, E.; Laugier, P.; Peyrin, F.; Padilla, F. Attenuation in trabecular bone: A comparison between numerical simulation and experimental results in human femur. *J. Acoust. Soc. Am.* **2007**, *122*, 2469–2475.

24. Muller, M; Moilanen, P.; Bossy, E.; Nicholson, P.; Kilappa, V.; Timonen, J.; Talmant, M.; Cheng, S.; Laugier, P. Comparison of three ultrasonic axial transmission methods for bone assessment. *Ultrasound Med. Biol.* **2005**, *31*, 633–642.

25. Dodd, S.P.; Cunningham, J.L.; Miles, A.W.; Gheduzzi, S.; Humphrey, V.F. Ultrasound transmission loss across transverse and oblique bone fractures: An *in vitro* study. *Ultrasound Med. Biol.* **2008**, *34*, 454–462.

26. Potsika, V.; Spiridon, I.; Protopappas, V.; *et al.* Computational study of the influence of callus porosity on ultrasound propagation in healing bones. In Proceedings of 36th Annual International Conference of the IEEE Engineering in Medicine and Biology Society, Chicago, IL, USA, 3–6 September 2014; pp. 636–638.

27. Potsika, V.T.; Grivas, K.N.; Protopappas, V.C.; Vavva, M.G.; Raum, K.; Rohrbach, D.; Polyzos, D.; Fotiadis, D.I. Application of an effective medium theory for modeling ultrasound wave propagation in healing long bones. *Ultrasonics* **2014**, *54*, 1219–1230.

28. Moilanen, P.; Nicholson, P.H.F.; Kilappa, V.; Cheng, S.; Timonen, J. Assessment of the cortical bone thickness using ultrasonic guided waves: modelling and *in vitro* study. *Ultrasound Med. Biol.* **2007**, *33*, 254–262.

29. Raum, K.; Grimal, Q.; Varga, P.; Barkmann, R.; Glüer, C.C.; Laugier, P. Ultrasound to Assess Bone Quality. *Curr. Osteoporos. Rep.* **2014**, *12*, 154–162.

30. Raum, K.; Leguerney, I.; Chandelier, F.; Talmant, M.; Saïed, A.; Peyrin, F.; Laugier, P. Site-matched assessment of structural and tissue properties of cortical bone using scanning acoustic microscopy and synchrotron radiation μCT.*Phys. Med. Biol.* **2006**, *51*, 733–746.

31. Potsika, V.T.; Gortsas, T.; Grivas, K.N.; Protopappas, V.C.; Vavva, M.G.; Raum, K.D.; Polyzos, D.; Fotiadis, D.I. Computational modeling of guided ultrasound wave propagation in healthy and osteoporotic bones. In ELEMBIO 6th National Conference, Patras, Greece, 2014.

32. Roschger, P.; Misof, B.; Paschalis, E.; Fratzl, P.; Klaushofer, K. Changes in the Degree of Mineralization with Osteoporosis and its Treatment. *Curr. Osteoporos. Rep.* **2014**, *12*, 338–350.

33. Cortet, B.; Boutry, N.; Dubois, P.; Legroux-Gérot, I.; Cotton, A.; Marchandise, X. Does quantitative ultrasound of bone reflect more bone mineral density than bone microarchitecture. *Calcif. Tissue Int.* **2004**, *74*, 60–67.

34. Sasso, M; Haïat, G.; Yamato, Y.; Naili, S.; Matsukawa, M. Frequency Dependence of Ultrasonic Attenuation in Bovine Cortical Bone: An In Vitro Study. *Ultrasound Med. Biol.* 2007, *33*, 1933–1942.

CHAPTER 2

Theoretical Analysis of SAW Propagation in 3C-SiC/c-AlN

Cinzia Caliendo

Istituto di Acustica e Sensori Corbino, IDASC-CNR, Via del Fosso del Cavaliere 100, Roma 00133, Italy

ABSTRACT

The anisotropic materials as the acoustic wave propagating medium introduce the dependence of the phase velocity on the wave propagation direction, as opposed to the isotropic counterparts; in addition, the profile of the particle displacement components can be quite different, depending on the crystal type and propagation direction. The propagation of surface and bulk acoustic waves (SAWs and BAWs) along the (001), (111) and (110) planes of cubic SiC crystals have been studied. For specific propagation directions in these planes, slight variations in the velocity of the elastic surface waves are found. It was observed that Rayleigh-type, generalized and pseudo-surface waves can propagate at specific directions, thus confirming how the anisotropic behavior of the bare SiC substrate modifies the existence and the field profile of the SAW that propagates at its free surface. Finally, the SAW propagation along AlN/SiC-based multilayered structures is studied for the three SiC planes, different AlN piezoelectric layer thicknesses and electrical boundary conditions.

KEYWORDS

3C-SiC; AlN; anisotropy; SAW; BAW

1. INTRODUCTION

Bulk acoustic waves (BAWs) propagating along an anisotropic medium show characteristics different as compared to the isotropic case. For higher symmetry crystals, if the wave propagation direction coincides with the principal axis direction, then three pure modes propagate having their polarization vectors parallel to the reference system axis: the longitudinal, shear horizontal and shear vertical bulk acoustic waves (LBAW, SHBAW and SVBAW). Nevertheless, for an arbitrary propagation direction, the polarization vector of the three bulk waves is neither necessarily parallel nor normal to the slowness vector: the *quasi* longitudinal and the two *quasi* shear waves show three displacement vector components. Surface acoustic waves (SAWs) propagating in isotropic media are elliptically polarized in the sagittal plane (the plane perpendicular to the free surface and containing the propagation vector): their phase velocity is unaffected by the propagation direction, and, consequently, the phase and group velocities coincide. On the contrary, the properties of the SAWs propagating along an anisotropic medium are mainly constrained by the crystallographic orientation of the free surface and by the direction of propagation. The SAWs show, in general, three displacement components, and the phase velocity depends on the propagation direction and does not coincide with the group velocity. Moreover, for specific propagation directions, the SAW, in its more general definition, does not exist and unconventional surface waves propagate, such as leaky or pseudo SAWs (PSAWs). Therefore, the anisotropic behavior of the medium affects the BAW characteristics, and considerably modifies the existence and the structure of the SAW that propagates at its free surface.

In the present paper, the propagation of BAWs and SAWs along the (001), (111) and (110) surfaces of bare SiC cubic crystals (3C-SiC) is studied. Specific directions were found suitable for the propagation of slow and fast *Rayleigh-like* waves, generalized SAWs, and pseudo-SAWs. Moreover, the SAW propagation along 3C-SiC/c-AlN-based multilayered structures is studied for propagation directions suitable for SAW device applications; different piezoelectric layer thicknesses and electrical boundary conditions were examined. This specific combination of materials was chosen for its suitability for applications in wireless communications and harsh environment sensing fields [1,2,3,4,5], such as extremes of pressure, temperature, shock, radiation and chemical attack. The piezoelectric AlN overlay material has excellent piezoelectric properties, low conductivity, and behaves elastically isotropic for surface waves propagating over the basal plane, provided that the AlN *c*-axis is perpendicular to the SiC substrate surface. The application of the largely-explored piezoelectric AlN film technology, together with the early-developed 3C-SiC film technology [6,7], have just produced some remarkable results [8,9,10], but, since the SiC technology is far from being mature, many more benefits will be attempted. *Bulk* SiC production is still in progress while SiC *film* on Si substrates are commercially available even if for limited thickness range. In 3C-SiC/c-AlN-based electroacoustic devices, the SiC anisotropy strongly affects the characteristics of the acoustic waves propagation rather than the c-AlN that is isotropic in the c plane.

Hence the propagation characteristics are influenced by the SiC thickness, cut and propagation direction. AlN, which shows the highest SAW velocity between the piezoelectric materials that can be grown in thin film form, has been successfully used in SAW devices operating to gigahertz frequencies. Thus, AlN, in combination with a fast substrate material such as SiC, creates new and exciting opportunities for miniaturizing, reducing power consumption, and improving the performance and functionality of many devices including sensors, actuators, radio frequency micro-electro-mechanical systems (RF MEMS) and optical arrays. Such 3C-SiC/c-AlN devices are robust, providing improved chemical stability, thermal stability, reproducibility and sensitivity compared to conventional polycrystalline piezoelectric MEMS.

2. BAW AND SAW PROPAGATION IN 3C-SIC

A mathematical model was built based on the partial differential equations system derived from the fundamental equations of motion in an elastic, lossless, homogeneous, anisotropic medium [11]. By solving the system of equations, the velocity of the bulk waves propagating along an arbitrary direction was obtained. Because of the anisotropy of the 3C-SiC, the eigenvalues and eigenvectors of the Christoffel's equation were affected by the arbitrary choice of the wave propagation direction. Thus, the phase velocities and the amplitudes of three acoustic modes were obtained for each propagation direction: one wave with longitudinal (or *quasi*-longitudinal) polarization and two waves with transverse (or *quasi*-shear vertical and *quasi*-shear horizontal) polarization [12]. A computer program was developed with Matlab software to determine the acoustic properties of the BAWs propagating along an arbitrary direction of an unbounded, homogeneous, lossless, linearly elastic anisotropic medium. The material constants of 3C-SiC substrates were those indicated in

[13]. The theoretical analysis of SAWs characteristics in 3C-SiC was accomplished by solving the continuum equations of motion, the strain-mechanical displacement relations, and the appropriate boundary conditions (traction free boundary conditions at the surface of the half space).

2.1. The (001) Plane of 3C-SiC

The phase velocity of the surface and bulk acoustic waves propagating in the (001) plane of 3C-SiC was calculated for different propagation directions, as shown in Figure 1.

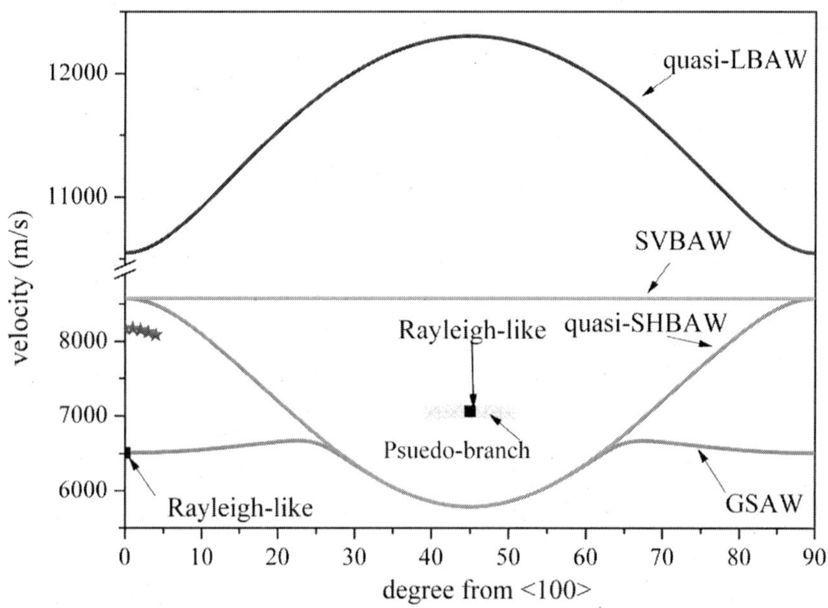

Figure 1. The angular dispersion of the BAWs and SAWs propagating in the (001) plane of 3C-SiC.

The SAWs and BAWs have the propagation vector lying in the (001) plane of SiC, in the direction indicated by the abscissa of the graph shown in Figure 1. At 0° from the <100> direction, as well as at 90°, the velocity of the two shear bulk waves coincides. The SVBAW is a pure mode polarized along the normal to the (001) plane for every propagation direction in the plane; the SHBAW and LBAW, except at 0°, 45° and 90°, are *quasi*-longitudinal and *quasi*-shear horizontal. With increasing the angle from the <100> direction, the SHBAW velocity decreases and becomes comparable to the velocity of the SAW, while the SVBAW velocity remains constant. For angles off x-axis greater than 45°, the SHBAW velocity increases and again coincides with the SVBAW velocity at 90°. The velocity of the LBAW increases for angles from 0° to 45°, and decreases for angles greater than 45° up to 90°.

At 0° and for few angles neighboring with this direction, the shear bulk wave satisfies the traction free boundary conditions: the wave travels at velocity $v = 8181$ m/s (the blue stars in Figure 1), and the particle displacement is in the sagittal plane, but it is not a surface wave since the wave amplitude does not decay with depth [12], as shown in Figure 2.

At 0° from the <100> direction, the SAW travelling at $v = 6510$ m/s is a *Rayleigh-like* wave: the two particle displacement components, the longitudinal and vertical components U_1 and U_3, show a damped oscillatory nature with depth [12], as shown in Figure 3. The *Rayleigh-like* mode polarization is elliptical with the ellipse lying in the sagittal plane and having the principal axis perpendicular to the free surface: U_1 and U_3, are in phase quadrature with each other, and the power flux is collinear with the propagation vector.

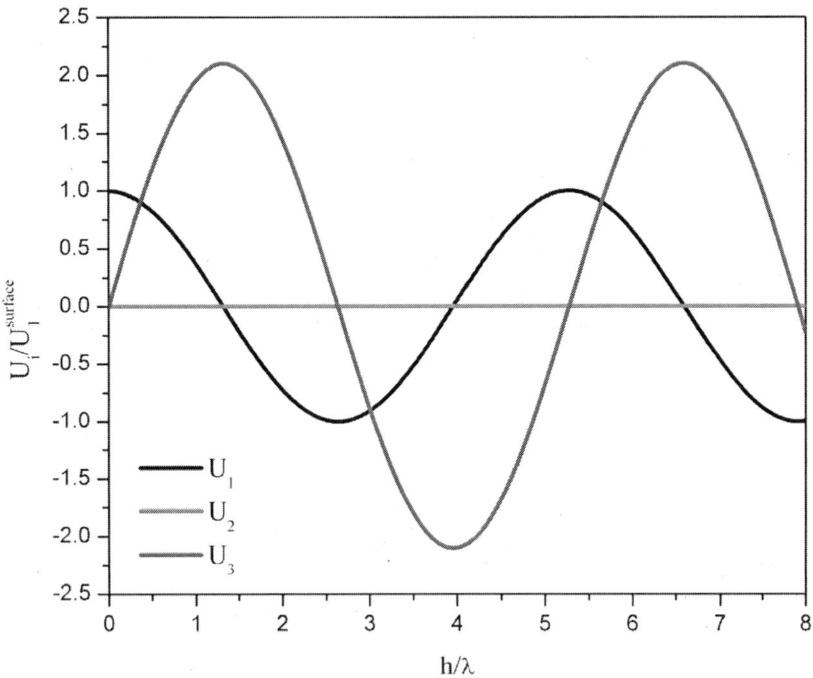

Figure 2. The normalized particle displacement components of the BAW that satisfies the traction free boundary conditions for the <100> direction in 3C-SiC(001) at velocity$v$ = 8181 m/s.

As the propagation vector is rotated from the <100> direction, the *Rayleigh-like* wave evolves to a generalized surface wave (GSAW) with three particle displacement components U_1, U_2 and U_3. When the propagation vector approaches approximately 33° off the <100> direction, the GSAW velocity becomes comparable to that of the *quasi*-SHBAW and the U_2component increases rapidly: from here on, the GSAW becomes leaky and degenerates in a bulk shear wave. At 45° off the <100> direction, there are two eigensolutions to the traction free boundary condition at the (001) plane of 3C-SiC. One solution is a *Rayleigh-like* wave travelling at velocity v = 7060 m/s higher than that of the slowest shear bulk wave (5785 m/s); this upper solution has the same characteristics of the *Rayleigh-like* wave solution at 0° and they both show a zero power flow angle (PFA) since the group and the phase velocity are collinear [12]. The lower solution at $v = 5785$ m/s is a wave with a dominant U_2 displacement component almost independent on the depth. Figure 4 shows the displacement components depth profile of the upper solution.

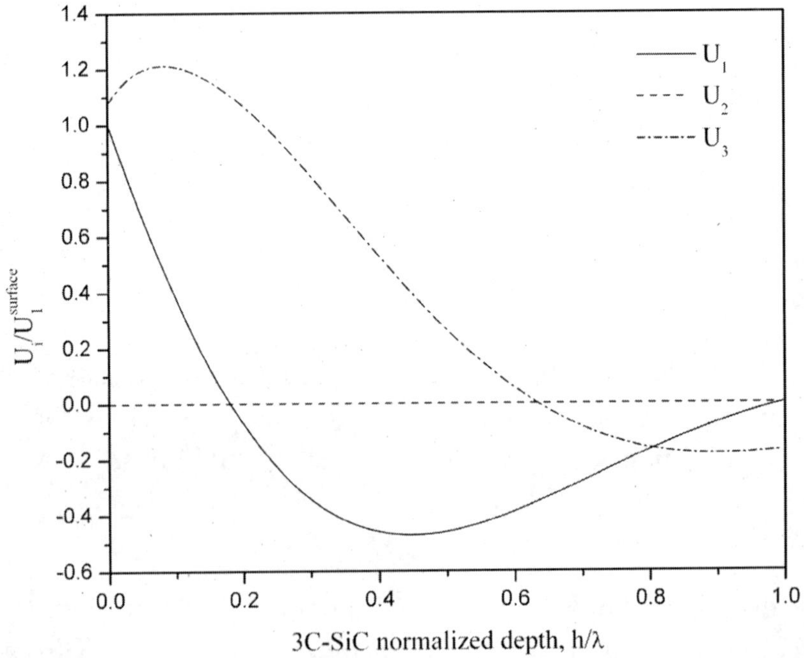

Figure 3. The particle displacement components of the *Rayleigh-like* wave propagating along the *x*- axis in 3C-SiC(001).

For angles neighboring with the upper solution, except at 45°, a generalized surface wave propagates having three displacement components that become appreciable at greater and greater depths, and hence the wave radiates energy into the bulk. These waves are called "pseudo SAW" (PSAW) [12]. As an example, Figure 5a shows the normalized particle displacement components of the PSAW propagating at 49° off the <100> axis: while the U_1 and U_3 components decay with the depth, the U_2 component does not decay and is responsible for the acoustic energy radiated into the bulk of the substrate. Figure 5b shows the PSAWs velocity and propagation loss per wavelength along the propagation direction *vs.* the angle off the <100> direction: as it can be seen, the propagation loss is zero at 45° off <100> where a SAW propagates, while it increases when the wave vector is rotated from the <110> direction.

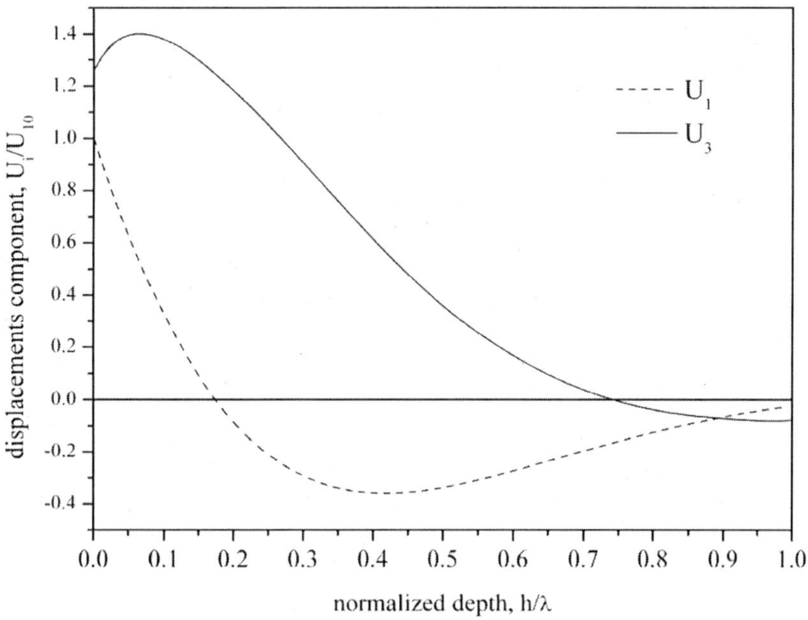

Figure 4. Variation of the normalized vertical and longitudinal displacement components with depth for the *Rayleigh-like* wave propagating along the <110> direction of the (001) plane of 3C-SiC at velocity $v = 7060$ m/s.

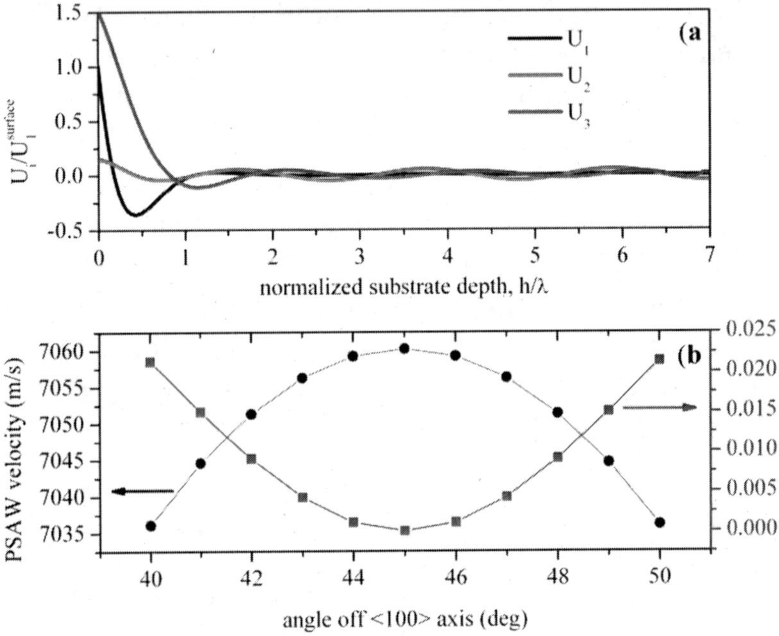

Figure 5. (**a**) Variation of the normalized shear and longitudinal displacement components with depth for the PSAW propagating at 49° off the <100> directions of the (001) plane of SiC; (**b**) the PSAWs velocity and propagation loss *vs.* the angle off the <100> direction of the (001) plane of SiC.

2.2. The (111) Plane of 3C-SiC

The SAWs and BAWs angular velocity dispersion curves in the (111) plane of 3C-SiC are shown in Figure 6; the curves have mirror symmetry about the 30° direction and hexagonal symmetry on this surface. At 0° from <110> direction, the three BAWs have equal phase and group velocity, as well at 60°, with zero PFA. The LBAW is pure while the two shear waves are*quasi*-SHBAW and *quasi*-SVBAW. With increasing the angle, the three BAWs become *quasi*-longitudinal and *quasi*-shear: their displacement components are mixed and their PFA is not zero. At 60° off, the

<110> direction the LBAW is pure as for 0°; the SHBAW is pure at 30° and 90°, but its phase velocity differs from the group velocity, unlike at 0°. The SVBAW is not a pure mode for any angle, but its phase velocity equals the group velocity at 0° and 60°.

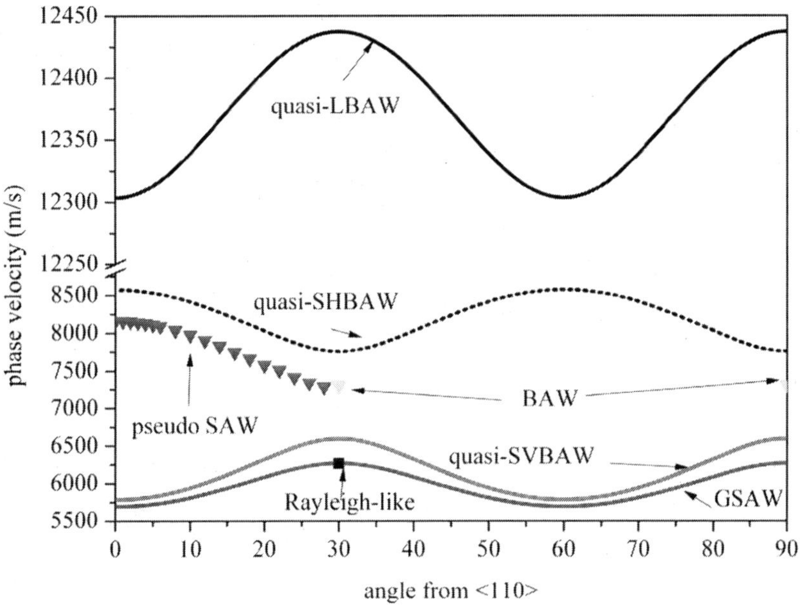

Figure 6. The angular dispersion curves of the SAWs and BAWs propagating along the (111) plane of 3C-SiC.

For SAW propagating along 0° (the <110> direction, that is normal to a plane of mirror symmetry) the surface displacement ellipse has its minor axis parallel to the direction of propagation while the other principal axis is not perpendicular to the free surface. The plane of the ellipse is rotated some 26° out of the sagittal plane on the free surface. The SAW velocity (5695 m/s) is only about 1.6% below the lowest BAW velocity (5785 m/s) in the same direction and has three displacement components. The penetration of this solution into the bulk of the material is larger than most solutions that are leaky at $h/\lambda = 1$: at one wavelength depth, the present solution shows $U_1/U_{1surface} = 0.037$, $U_2/U_{1surface} = 0.79$ and $U_3/U_{1surface} = 1.1$, as shown in Figure 7a.

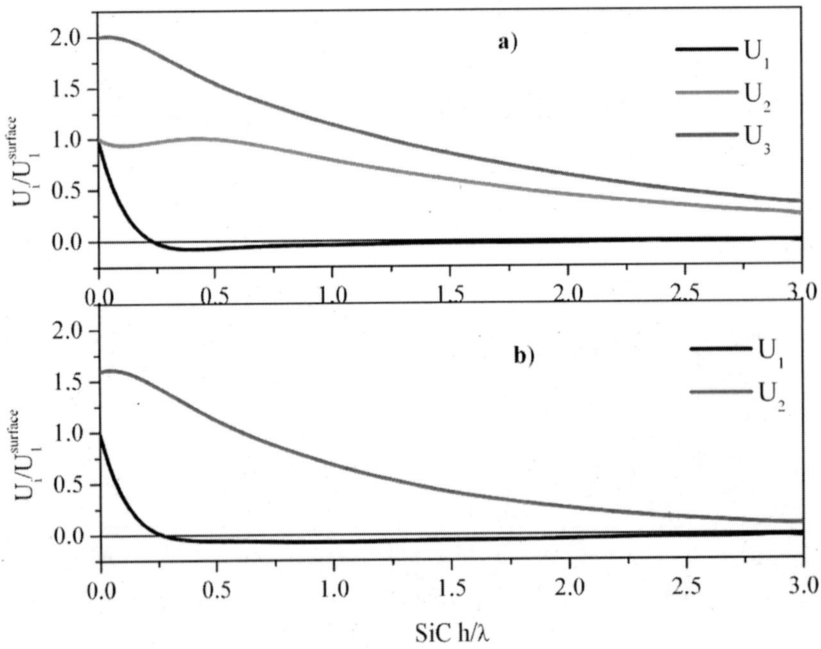

Figure 7. Variation of the normalized shear and longitudinal displacement components with depth for the SAW propagating along the <110> direction (**a**) and 30° off the <110> direction (**b**) of the (111) plane of 3C-SiC.

Between 0° and 30°, the surface acoustic wave is a GSAW with three displacement components. At 30° in the (111) plane, the surface displacement ellipse lies in the sagittal plane as for a *Rayleigh-like* wave ($U_2 = 0$ and U_1 and U_3 have oscillatory damped behavior, as shown in Figure 7b), but the ellipse major axis is not vertical (the angle between U_1 and U_3 is not 90° but 105° from the U_1 direction, *i.e.*, the propagation direction).

Between 0° and 30°, PSAW propagates at higher velocity than that of the SAW, having three displacement components that are not confined to the vicinity of the surface: the penetration depth of this waves increases gradually with increasing the propagation angle off the <110>, and the PSAW radiates energy into the bulk. Figure 8a–d show the field profile of the PSAW travelling at 0°, 10°, 20° and 25° off the <110> direction, at velocity equal to 8161.75, 7972.96, 7523.64 and 7370.43, and with propagation loss equal to 0.017, 0.26, 0.54 and 1.08 dB per wavelength, respectively. As it can be seen, with increasing the angle from the <110> direction, the penetration into the bulk gradually increases. At 30° (as well as at 90°), the pseudo SAW degenerates in a SHBAW having only one displacement component (U_2).

2.3. The (110) Plane of 3C-SiC

Figure 9 shows the phase velocity angular dispersion curves of the BAWs and SAWs propagating in the (110) plane of 3C-SiC. At 0° from the <001> direction in the (110) plane, as well as at 180°, the two bulk shear waves travel at the same velocity (8577 m/s): at this specific direction, the phase and group velocities of the three BAWs coincide. With increasing the angle off the <001> direction, the longitudinal wave becomes *quasi*-longitudinal and becomes again a pure mode at approx. 51° up to 90° off <100>. The *quasi*-SHBAW, having approximately mirror symmetry about 45°, has its particle displacement components in the (110) plane at all angles, and is a pure mode in the range from ~51° to 58° and at 90°, whereas the SVBAW has its displacement component perpendicular to this plane at all angles.

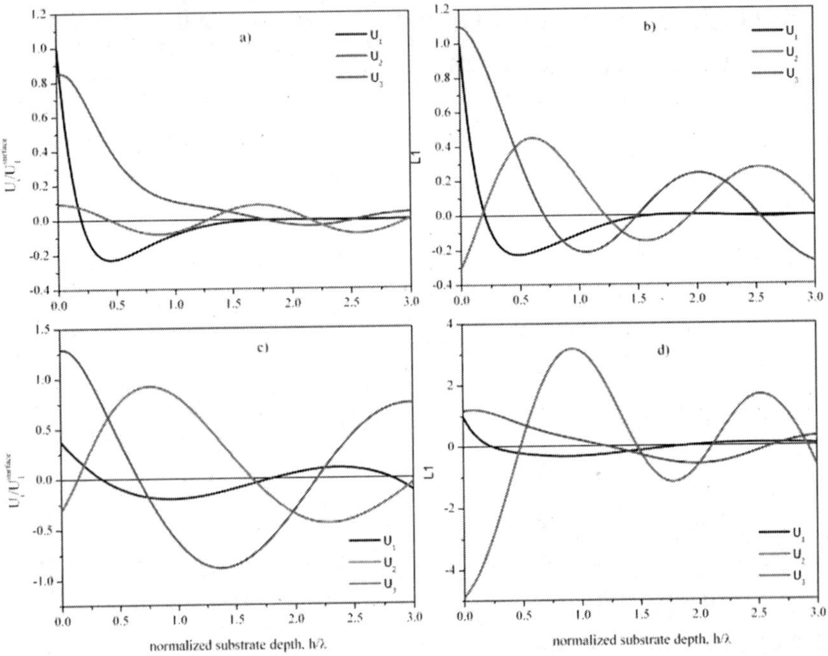

Figure 8. The field profile of the pseudo SAW travelling at: 0°(**a**); 10 (**b**); 20° (**c**) and 25° (**d**) off the <110> direction of the (111) plane of 3C-SiC.

Along the <001> direction of the (110) plane, a Rayleigh-like mode propagates with two displacement components U_1 and U_3 vanishing with depth in an oscillatory way, as shown in Figure 10a. Along the <110> direction, a *simple Rayleigh* wave propagates [12] with two displacement components U_1 and U_3, slowly damping into the bulk, as shown in Figure 10b.

For propagation directions between <001> and <110>, a GSAW propagates with three displacement components U_1, $U_2$and U_3. The U_2 component amplitude increases from 0° to 45° off <001> because of the approaching of the GSAW and SHBAW velocities, while it decreases from 45° to 90° with the increasing of the velocity difference between the GSAW and SHBAW. The U_2

component *vs.* the SiC normalized depth is shown in Figure 11, for different propagation directions in the (110) plane.

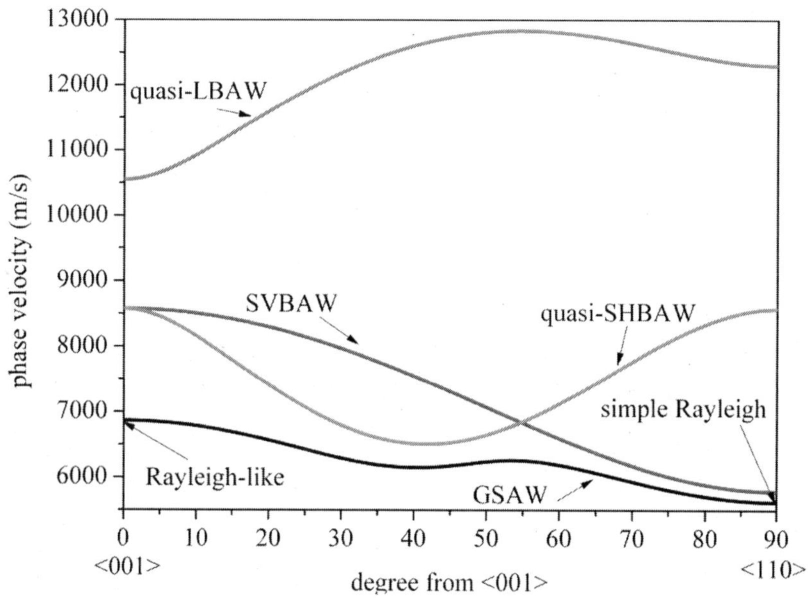

Figure 9. The phase velocity angular dispersion curves of the BAWs and SAWs propagating in the (110) plane of 3C-SiC.

No degeneration of SAW into a BAW at any angle was observed, and no pseudo surface-branch appeared. It is worth noting that the SHBAW satisfies the traction free boundary conditions for both the <001> and <110> directions, even though these solutions do not represent a surface wave.

3. ACOUSTIC WAVE PROPAGATION IN 3C-SIC/C-ALN

In a film/substrate structure, the acoustic wave propagation properties become dispersive with respect to the thickness of the layer. As a result, in the SiC/AlN multilayer structure, the SAW propagation properties would be dispersive with respect to the AlN film thickness normalized to the wavelength. Moreover, the phase velocity dispersion curves of SAW propagating along different directions in different planes of the SiC substrate covered by an AlN layer are not identical due to the cubic crystal system of the SiC half-space.

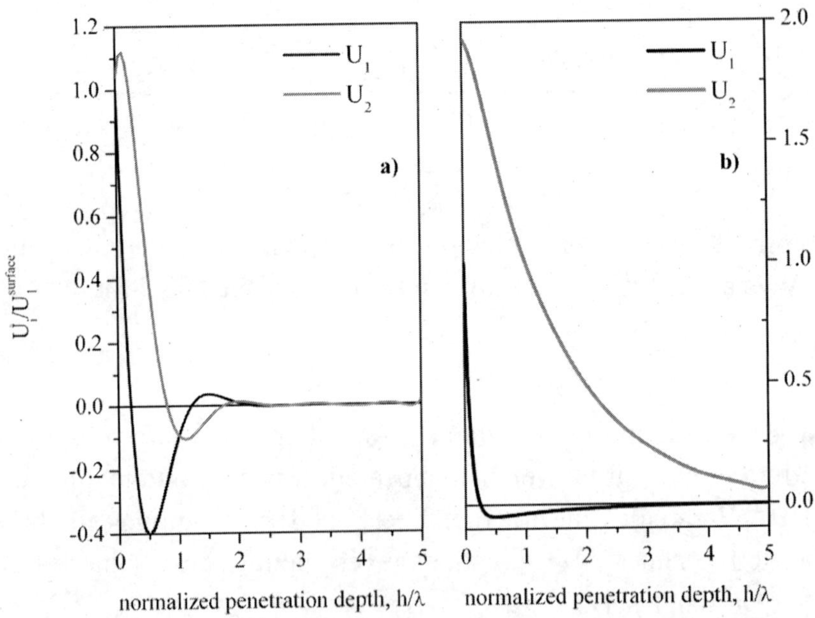

Figure 10. Variation of the normalized shear and longitudinal displacement components with depth of the *Rayleigh-like* wave propagating along <001>(110)SiC (**a**) and of the *simple Rayleigh* wave propagating along <110>(110)SiC (**b**).

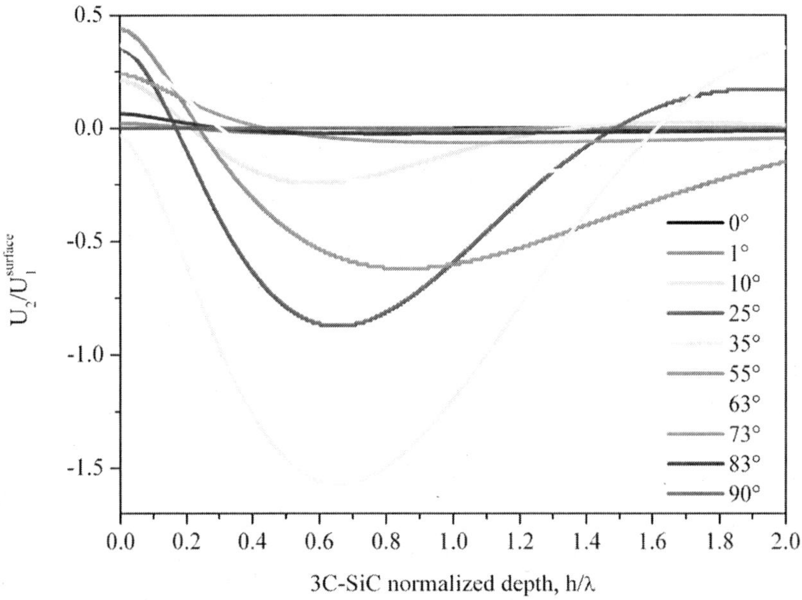

Figure 11. Variation of the normalized shear horizontal displacement component *vs.*the SiC depth, being the angle off the <001> axis the running parameter.

The phase velocity dispersion curves of SAW propagating along the (001), (111) and (110) planes of 3C-SiC substrate covered by a piezoelectric c-AlN layer was investigated for different propagation directions and electrical boundary conditions. The electromechanical coupling constant $K^2 = 2 \cdot [(v^f - v^m)/v^f]$, the percentage difference of the velocity between a free surface (v^f) and surface coated with an infinitesimally thin perfect conductor (v^m), is used as direct estimate of surface-wave coupling to interdigital transducers [14,15]. The SAW devices with four types of transducer configurations on the SiC/AlN multilayer structure are studied: the typologies called SFT (substrate/film/transducer) and SMFT (substrate/metal/film/transducer) represent a pair of transmitting and receiving interdigital transducers (IDTs) on the top AlN surface combined with and without a metallized interface in the SiC/AlN structure; the typologies called STF (substrate/transducers/film) and STFM (substrate/ transducer/ film/metal) represent a pair of IDTs in the SiC/AlN interface joined with a unmetalized and a metalized top AlN surface. In the theoretical calculation, the AlN thin film is treated as a hexagonal crystal system, and the SiC substrate is treated as a cubic crystal system whose thickness is assumed to be infinite. The velocity calculations were performed under the hypothesis of lossless materials, assuming the single crystal AlN and 3C-SiC material constants available in the literature [16,17].

3.1. Surface Acoustic Waves in 3C-SiC(001)/c-AlN

When the 3C-SiC half-space is covered by a thin c-axis oriented AlN layer, the surface wave velocity along the (001) plane of the SiC/AlN becomes dispersive. Both the SiC(001)<100>/c-AlN and SiC(001)<110>/c-AlN substrate are a combination of*slow* film on a *fast* substrate as the SAW velocity of AlN is lower than that of the substrate. Since the SVBAW in the film is slower than the SVBAW in the substrate, there exist higher Rayleigh-type modes having a low frequency cut off at which the phase velocity is equal to the substrate SVBAW velocity; with increasing the layer thickness, the velocity of these higher order modes asymptotically reaches the SVBAW velocity in the layer. Figure 12 shows the dispersion curves of the *first*Rayleigh-like mode (with two displacement components, U_1 and U_3) propagating along the <100> and <110> propagation directions. For vanishingly AlN thin layer thickness ($h/\lambda << 1$), the Rayleigh-like mode velocity equals the *Rayleigh-like* mode velocity in the bare SiC substrate. When the layer thickness becomes large compared to the wavelength ($h/\lambda > 1$), the wave velocity decreases and asymptotically reaches the SAW velocity appropriate to a free surface of the layer material (5607 m/s).

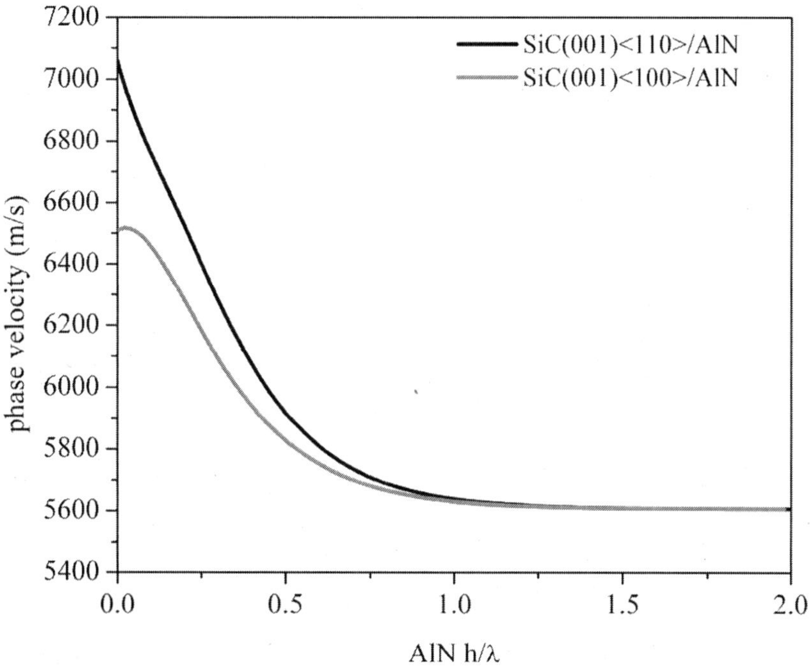

Figure 12. The phase velocity dispersion curves of the *Rayleigh-like* waves propagating along 3C-SiC(001)<100>/c-AlN and 3C-SiC(001)<110>/c-AlN structures.

Figure 13a,b show the K^2 dispersion curves of the four coupling configurations for the SAW propagating along the <100> and <110> directions of 3C-SiC(001)/AlN.

As it can be seen from Figure 13, the K^2 dispersion curves are strongly affected by the electrical boundary conditions that influence the potential depth profile. For acoustic waves propagating along layered structures, the achievable K^2 can be larger than that of AlN (0.3%) considered as a semi-infinite substrate, due to an increased acoustic energy confinement. Between the four coupling structures, the STFM and STF structures exhibit the highest K^2 value for an AlN normalized thickness equal to about 0.5.

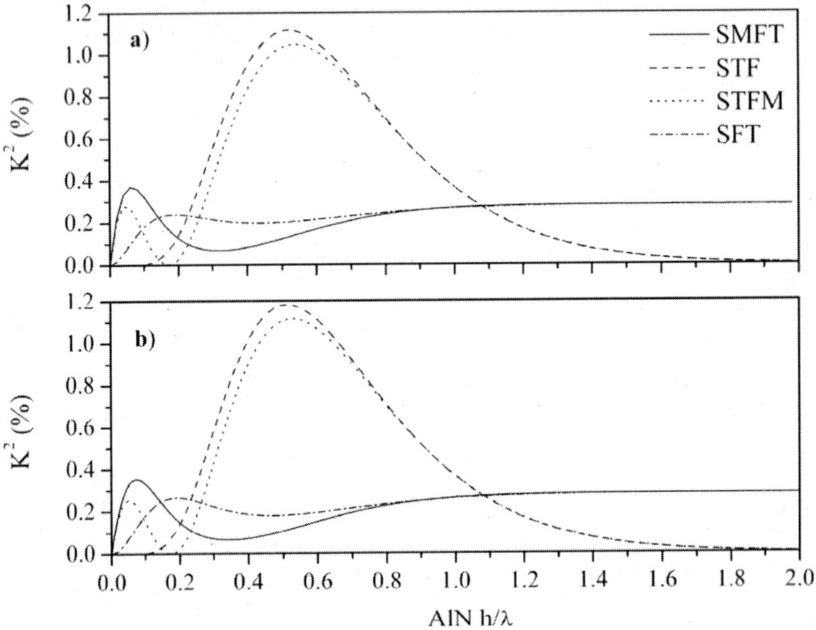

Figure 13. The K^2 dispersion curves of the *Rayleigh-like* waves propagating along the <100> (**a**) and <110> (**b**) directions of the four 3C-SiC(001)/AlN-based coupling configurations.

3.2. Surface Acoustic Waves in 3C-SiC(111)/c-AlN

For propagation of SAW at 0° and 30° off the <110> axis, the 3C-SiC/AlN configuration is a *slow on fast* configuration as the SAW velocity in SiC is higher than that in c-AlN. Figure 14 shows the velocity and K^2 dispersion curves of SiC(111)/AlN for propagation along the <110> axis. No higher order modes propagate since the *quasi*-SVBAW velocity in SiC (5784 m/s) is much lower than that of the SVBAW in the AlN (6089 m/s).

Figure 15 shows the velocity and K^2 dispersion curves of SiC(111)/AlN for propagation along 30° off <110> axis. The structure sustains the propagation of higher order modes since the *quasi*-SVBAW velocity in SiC (6597 m/s) is much lower than that of the SVBAW in the AlN (6089 m/s).

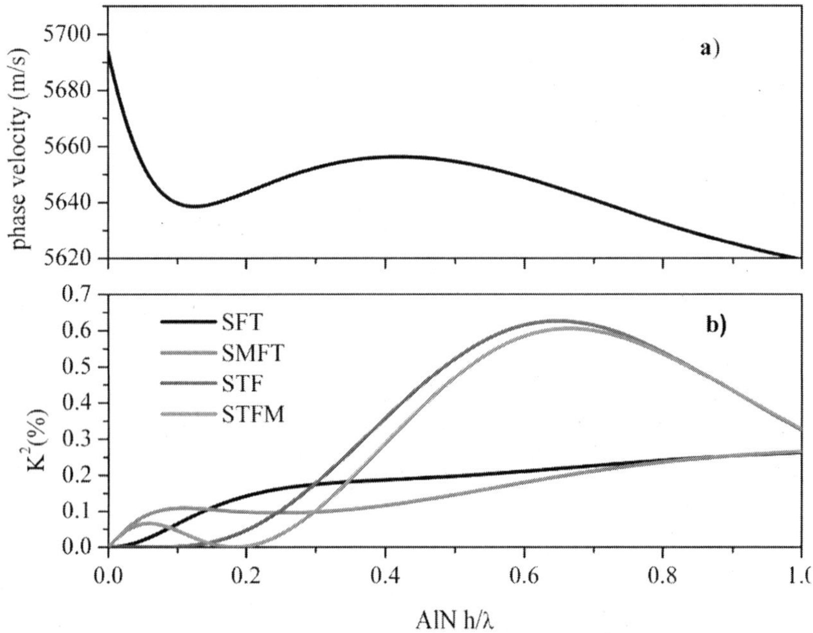

Figure 14. The phase velocity (**a**) and K^2 (**b**) dispersion curves for SAW propagating along the <110> direction of the (111) SiC/AlN substrate.

3.3. Surface Acoustic Waves in 3C-SiC(110)/c-AlN

The phase velocity and K^2 dispersion curves for the fundamental SAW propagating along the <001>(110)SiC/AlN *vs.* the AlN normalized thickness h/λ are shown in Figure 16a,b.

Such configuration sustains the propagation of higher order modes since the SVBAW velocity of the substrate is higher than that of the AlN layer. The phase velocity and K^2 dispersion curves for the

SAW propagating along <110>(110) SiC/AlN*vs.* the normalized AlN thickness h/λ are shown in Figure 17a,b.

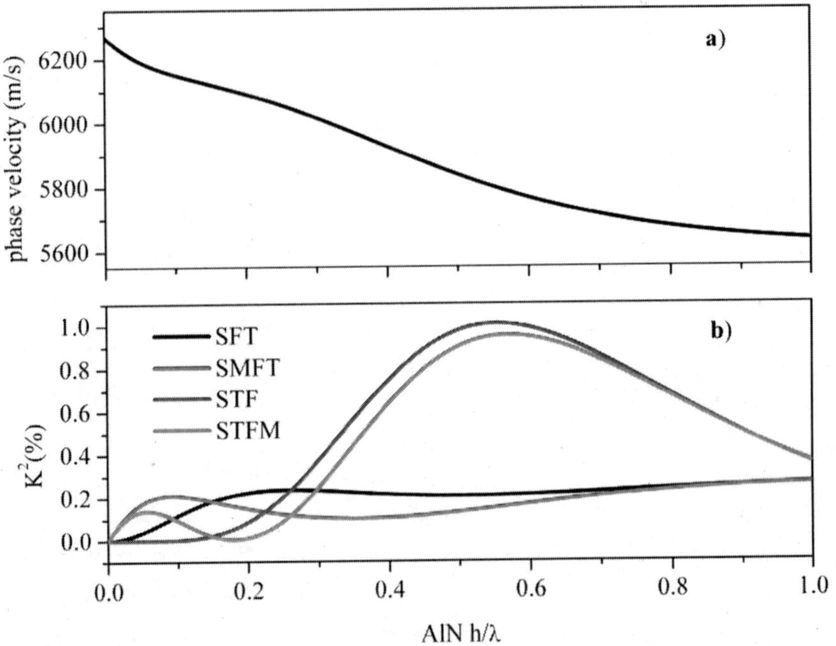

Figure 15. The phase velocity (**a**) and K^2 (**b**) dispersion curves for SAW propagating along 30° off the <110> direction of the (111) SiC/AlN substrate.

In this case, no higher order modes exist since the SVBAW velocity in AlN is much higher than that in SiC.

4. CONCLUSIONS

The SAW propagation properties through the 3C-SiC/AlN composite structure were theoretically analyzed for different SiC planes, AlN layer thicknesses, acoustic wave propagation directions and electrical boundary conditions. The calculations performed demonstrate the potential of AlN thin films grown on epitaxial 3C-SiC layers to create layered SAW devices with high phase velocity and large electromechanical coupling coefficient. The simulation results show that SAW propagating along the <110> direction of the (001) plane has the highest phase velocity among the other directions studied: for AlN thickness of about $0.1\cdot\lambda$, the velocity is about 7000 m/s, with a K^2 ~0.4% for the SMFT configuration; while for an AlN thickness of about $0.5\cdot\lambda$, the K^2 is about 1.2% and the velocity is about 6000 m/s for the STF configuration. The results conclude that the 3C–SiC/AlN multilayer structure possesses high surface wave velocity and satisfactorily large coupling coefficient, which enables high-frequency layered SAW devices applicable to telecommunications and sensing fields applications.

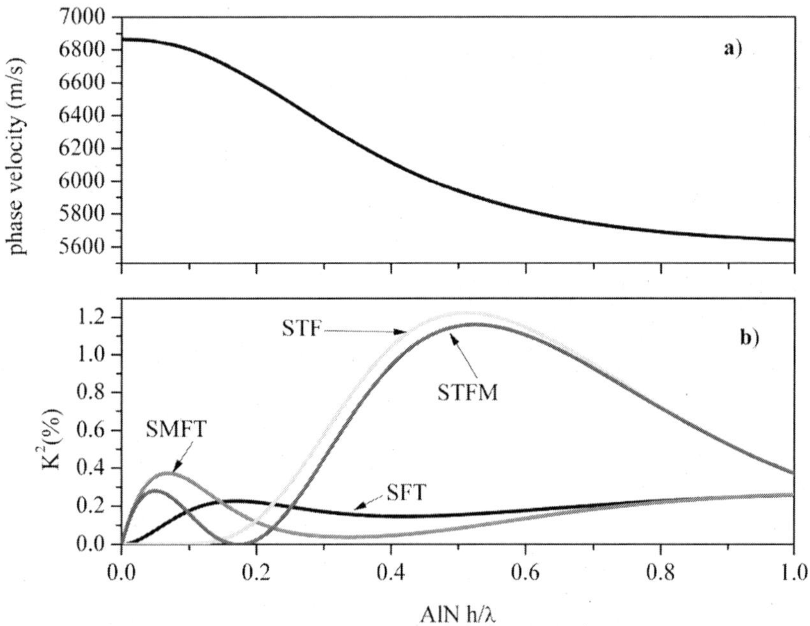

Figure 16. The phase velocity (**a**) and K^2 (**b**) dispersion curves for SAW propagating along the <001> direction of the (110) SiC/AlN substrate.

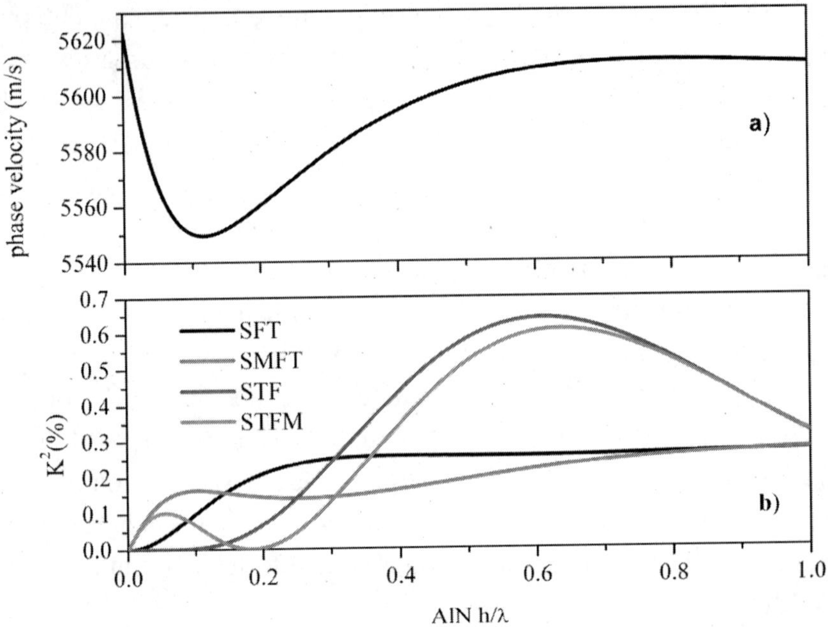

Figure 17. The phase velocity (**a**) and K^2 (**b**) dispersion curves for SAW propagating along the <110> direction of the (110) SiC/AlN substrate.

ACKNOWLEDGMENTS

The author wishes to thank the Institute for Photonics and Nanotechnology, Rome Unit, who hosted her during the research period.

CONFLICTS OF INTEREST

The author declares no conflict of interest.

REFERENCES

1. Chung, G.-S.; Jeong, J.-M. Fabrication of micro heaters on polycrystalline 3C-SiC suspended membranes for gas sensors and their characteristics. *Microelectron. Eng.* **2010**, *87*, 2348–2352.

2. Doppalapudi, D.; Mlcak, R.; Chan, J.; Tuller, H.; Bhattacharya, A.; Moustakas, T. MBE Grown AlN Films on SiC for Piezoelectric MEMS Sensors. In Proceedings of the MRS Symposium Y—GaN and Related Alloys, Boston, MA, USA, 1–5 December 2003; pp. 403–408.

3. Lin, C.; Chen, Y.; Felmetsger, V.V.; Lien, W.; Riekkinen, T.; Senesky, D.G.; Pisano, A.P. Surface acoustic wave devices on AlN/3C–SiC/Si multilayer structures. *J. Micromech. Microeng.* **2013**, *23*, 025019.

4. Lin, C.; Chen, Y.; Pisano, A.P. Theoretical investigation of Lamb wave characteristics in AlN/3C–SiC composite membranes. *Appl. Phys. Lett.* **2010**, *97*, 193506.

5. Caliendo, C. Theoretical investigation of Lamb wave A_0 mode in thin SiC/AlN membranes for sensing application in liquid media. *Sens. Actuators B Chem.* **2013**, *179*, 287–292.

6. Liu, F.; Carraro, C.; Chu, J.; Maboudian, R. Residual stress characterization of polycrystalline 3C-SiC films on Si(100) deposited, from methylsilane. *J. Appl. Phys.* **2009**, *106*, 013 505.

7. Roper, C.S.; Howe, R.T.; Maboudian, R. Stress control of polycrystalline 3C-SiC films in a large-scale LPCVD reactor using 1,3-disilabutane and dichlorosilane as precursors. *J. Micromech.Microeng.* **2006**, *16*, 2736–2739.

8. Lin, C.; Lien, W.; Felmetsger, V.V.; Hopcroft, M.A.; Senesky, D.G.; Pisano, A.P. AlN thin films grown on epitaxial 3C–SiC (100) for piezoelectric resonant devices. *Appl. Phys. Lett.* **2010**, *97*, 141907.

9. Lin, C.; Chen, Y.; Felmetsger, V.; Senesky, D.; Pisano, A. AlN/3C-SiC composite plate enabling high frequency and high Q micromechanical resonators. *Adv. Mater.* **2012**, *24*, 2722–2727.

10. Caliendo, C. Theoretical investigation of high velocity, temperature compensated Rayleigh waves along AlN/SiC substrates for high sensitivity mass sensors. *Appl. Phys. Lett.* **2012**, *100*, 021905.

11. Adler, E.L. SAW and pseudo-SAW properties using matrix methods. *IEEE Trans. Ultrason. Ferroelect. Freq. Control* **1994**, *41*, 699–705.

12. Farnell, G.W. Properties of elastic surface waves. In *Physical Acoustics, Principles and Methods*; Mason, W.P., Thurston, R.N., Eds.; Academic Press: New York, NY, USA, 1970; Volume VI, Chapter 4; p. 121.

13. Levinshtein, M.E.; Rumyantsev, S.L.; Shur, M.S. *Properties of Advanced Semiconductor Materials: GaN, AlN, InN, BN, SiC, SiGe*; Wiley: New York, NY, USA, 2001.

14. Collins, J.H.; Gerard, H.M.; Shaw, H.J. High performance lithium niobate acoustic surface wave transducers and delay lines. *Appl. Phys. Lett.* **1968**, *13*, 312–313.

15. Ingebrigtsen, K.A. Surface waves in piezoelectrics. *J. Appl. Phys.* **1969**, *40*, 2681–2686.

16. Tsubuochi, K.; Sugai, K.; Mikoshiba, N. AlN Material Constants Evaluation and SAW Properties on AlN/Al_2O_3 and

AlN/Si. In Proceedings of the IEEE Ultrasonics Symposium, Chicago, IL, USA, 14–16 October 1981; pp. 375–380.

17. Hellwege, K.-H.; Hellwege, A.M. *Landolt-Börnstein, Numerical Data and Functional Relationships in Science and Technology, New Series, Group III*; Springer: Berlin, Germany, 1979.

CHAPTER 3

Dynamic Characterization of Cohesive Material Based on Wave Velocity Measurements

Wojciech Sas [1,*], Katarzyna Gabryś [1], Emil Soból [2] and Alojzy Szymański [2]

[1] Water Centre Laboratory, Faculty of Civil and Environmental Engineering, Warsaw University of Life Sciences, 02-787 Warsaw, Poland

[2] Department of Geotechnical Engineering, Faculty of Civil and Environmental Engineering, Warsaw University of Life Sciences, 02-787 Warsaw, Poland

ABSTRACT

The paper presents a description of the dynamic properties of cohesive material, namely silty clays, obtained by using one of the applied seismology methods, the bender elements technique. The authors' aim was to present the dynamics of a porous medium, in particular an extremely important passage of seismic waves that travel through the bulk of a medium. Nowadays, the application of the bender element (BE) technique to measure, e.g., small strain shear stiffness of soils in the laboratory is well recognized, since it allows for reliable and relatively economical shear wave velocity measurements during various laboratory experiments. However, the accurate estimation of arrival time during BE tests is in many cases unclear. Two different interpretation procedures (from the time domain) of BE tests in order to measure travel times of waves were examined. Those values were then used to calculate shear and compression wave velocities and elastic moduli. Results showed that the dynamic parameters obtained by the start-to-start method were always slightly larger (up to about 20%) than those obtained using the peak-to-peak one. It was found that the peak-to-peak method led to more scattered results in comparison to the start-to-start method. Moreover, the influence of the excitation frequency, the mean effective stress and the unloading process on the dynamic properties of the tested material was studied. In addition, the obtained results highlighted the importance of initial signal frequency and the necessity to choose an appropriate range of frequencies to measure the shear wave velocity in clayey soils.

KEYWORDS

cohesive material; various elastic moduli; seismic wave propagation velocity; laboratory tests; bender element technique

1. INTRODUCTION

The determination of seismic velocities, as well as the elasticity modulus and structural properties of porous materials plays an extremely important role in the development of various engineering projects [1]. The small strain and elastic moduli, such as the shear modulus (G_{max}) and Young's modulus (E_{max}), are key parameters during, for example, the site response analysis of an earthquake, the design of machine foundations and soil dynamics problems [2]. Burland *et al.* [3] even emphasized the significance of G_{max} in static deformation analysis of geotechnical problems. Usually, the above-mentioned elastic modulus of soil deposits is measured *in situ* by means of so-called seismic exploration [4]. In order to evaluate the shear modulus (G) or Young's modulus (E) values in the laboratory, triaxial or torsional shear testing machines, which employ what is generally known as the static loading methods [5], are applied. Other kinds of evaluation methods, commonly referred to as vibration test methods, are those relying on applying wave motions to the test specimens and then observing their behavior during the resonance, including free oscillation time. A great example is a resonant column apparatus. Apart from this, there are also pulse transmission techniques, which include the ultrasonic pulse test, bender element (BE), *etc.* They allow calculating G_{max} at small strains on the basis of the wave velocity [6].

Some of the laboratory tests and field studies, such as cross-hole seismic [7,8], down-hole seismic [9], suspension logging [10], seismic cone [11], seismic flat dilatometer [12] and spectral analysis of seismic waves [13], are indirect tests [14]. Compared to direct tests, the indirect tests enable measuring quantities different from the more desired ones and relate them to each other through mathematical relationships. Seismic techniques are classified as indirect testing methods for small strain stiffness [7]. They yield profiles of wave propagation velocity. When we assume that the behavior of the material is linear-elastic, this means that elastic stiffness relates to wave propagation velocity following the equations below:

$$\upsilon p = \sqrt{\frac{\lambda + 2G}{\rho}} \tag{1}$$

$$\upsilon_S = \sqrt{\frac{G}{\rho}} \tag{2}$$

where υ_P is the propagation velocity of pressure, which might as well be called a primary (P-) wave velocity, υ_S is the propagation velocity of shear or the secondary (S-) wave and λ and G are Lame's constants (another term for G is the shear modulus). For engineers, it is more convenient to express Lame's constant as a function of Young's modulus and Poisson's ratio in accordance with these equations:

$$G = \frac{E}{2(1+\nu)} \tag{3}$$

$$\lambda = \frac{\nu E}{(1+\nu)(1-2\nu)} \tag{4}$$

Figure 1 illustrates the particle motion in P- and S-waves.

Over the years, bender elements have become increasingly popular in order to determine the S-wave velocity [9,10] through experiments. Nowadays, they are commonly available [11,12]. Bender elements consist of two sheets produced from piezoceramic plates, which are rigidly bonded, to a center brass shim or stainless steel plate. Because of their piezoelectric properties, they are capable of converting mechanical excitation into electrical output and *vice versa*. When excited by an input voltage, the bender element changes its shape, which is accompanied by mechanical excitation. Because of this fact, it acts as a signal transmitter. Being subjected to mechanical excitation causes the emission of an electrical output by another bender element, which is, in this way, acting as a signal receiver [9]. The first person to use piezoceramic elements bonded to shear plates was probably Lawrence [13,14]. His idea was to measure the velocity of the shear wave in materials, such as clay and sand. In 1978, Shirley [15] proposed the introduction of piezoceramic bender elements to determine S-wave velocity in the laboratory. Other scientists, like Dyvik and Madshus [16], described the incorporation of BE for geotechnical laboratory testing in the 1980s.

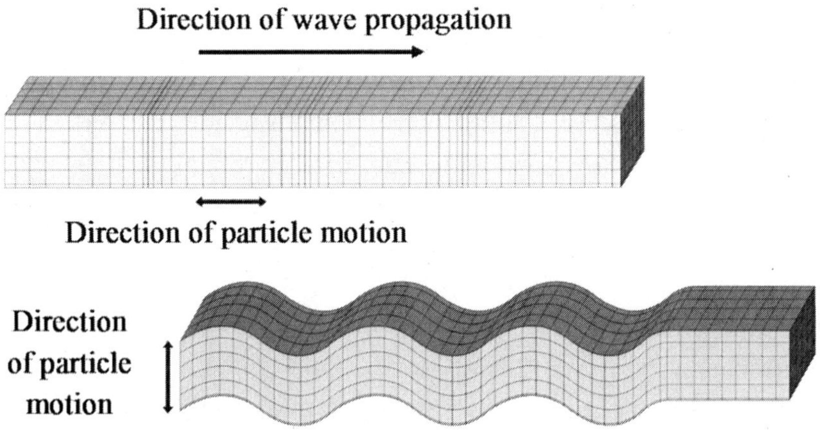

Figure 1. P-wave (top) and S-wave (bottom) particle motion. The particle motion in P-waves is longitudinal, while in S-waves, it is transverse. The particle motion vector in the plane perpendicular to the direction of propagation is referred to as polarization (only in the case of S-waves) [8].

The appeal of the BE technique lies in its apparent simplicity: one of the transducers is excited at one end of a specimen by a single pulse excitation, and the second, located at the other end of a specimen, is receiving it. The time required for this process can be simply read off from an oscilloscope. On the basis of this value, shear wave velocity can be obtained [17]. Therefore, BEs are an inexpensive and versatile solution, which can be used for laboratory seismic measurement. Furthermore, their capabilities of monitoring the process of stiffening with effective stresses, namely cementation, load stabilization, curing or consolidation, are particularly appealing [18]. However, there is a critical drawback connected with a BE test that contributes to many errors, namely the determination of the travel time. A number of researchers, including Viggiani and Atkinson (1995), attempted to use various methods in order to find the arrival time in order to reduce the degree of subjectivity [19]. In 2005, Lee and Santamarina [20], as well as Leong *et al.* [21] reported a method of travel time determination, by performing a bender element test. Nevertheless, there are still some differences between the first arrival of the S-wave and the preferred input wave. Therefore, the BE technique still needs to be verified with respect to how reliable it is, when it comes to calculation of the arrival time [2,22].

The intention of the authors of this paper was to recognize the dynamic characteristics of the cohesive medium through the performance of the BE tests on this material, *i.e.*, clayey soil, obtained from the area of Warsaw, the capital of Poland, at various effective confining pressures under saturated conditions. Two methods, namely peak-to-peak and start-to-start, were used to determine the travel time using the BE technique. The υ_S was calculated on this basis. Subsequently, shear modulus (G) was obtained using the value of shear wave velocity. Additionally, the authors used piezoceramic elements to determine compression wave velocity (υ_P). Furthermore, they tried to evaluate the values of dynamic Poisson's ratio (ν_d). The dynamic Poisson's ratio of soil deposits is a matter that has attracted rather little attention so far [23]. Nevertheless, it significantly affects stresses, strains, wave propagation characteristics in a mass of soils and other important types of soil deposit behavior during dynamic excitation [4]. In this paper, ν_d was calculated from the results of measurements of longitudinal and shear wave velocities in seismic exploration and then used to establish Young's modulus (E). The authors also studied the possible impact of unloading on the received values of elastic modulus.

2. METHODS AND MATERIALS

2.1. Test Equipment

For the experiments, the authors employed a Stokoe fixed-free type of resonant column, manufactured in 2009 by the British company GDS Instruments Ltd, with its office in Hook, Hampshire, U.K. A detailed description of this device is contained in the authors' other publications [24,25] and will not be discussed in this paper. The BEs, produced as well by GDS Instruments Ltd in 2009, were installed in the above-mentioned resonant column apparatus. The GDS bender elements are made from piezoelectric ceramic bimorphs. Two sheets are bonded together with a metal shim in between them. Excitation voltage is used to produce a displacement in the source transducer, resulting in a wave being sent through the sample. This wave generates a displacement in the receiver, which induces a voltage that can be measured.

The GDS bender elements come in the form of an insert that can be mounted in a top cap or a pedestal (Figure 2). This makes them easy to replace and able to be used with a wide range of triaxial cells. The insert for the base pedestal is made of stainless steel, while the one for the top cap is made of titanium. This reduces the weight by half and minimizes the axial load caused by the top cap, which is being applied to the sample. The inserts are embedded in a modified Perspex top cap and stainless steel base pedestal. At the same time, the top cap and the base pedestal are mounted on the specimen in the same way, as during a conventional triaxial test [26].

The S-wave source comprises two piezoceramic strips both polarized in the same direction. The application of an excitation voltage induces the extension of one strip and the contraction of the other, causing the strip to bend. Turning the excitation voltage in the opposite direction contributes to strips bending reversely. The two strips in the receiver are polarized in opposite directions.

The P-wave source consists of two piezoceramic strips polarized conversely. When an excitation voltage is applied, strips extend simultaneously. Reversion of the excitation voltage causes the strips to contract [26].

Figure 2. Bender element inserts mounted in a top cap and base pedestal.

In Figure 3, the BE arrangement adopted for the presented experiments is shown. Elements are manufactured to allow both S- and P-wave testing to be performed (in opposite propagation directions) on the same sample. The software switches input gain levels (of the received signal), sets the level of the output signal voltage and controls switching between P- and S-wave modes for our combined wave type elements [26]. In the current research, bender elements were used to obtain υ_{VH}velocities, propagating in the vertical plane (index *V*), polarized in the horizontal (index *H*).

When performing a bender element test, one of its most important aspects is the phase orientation of the elements. The authors always checked the relationship between the received signals with respect to the source signal. The desirable orientation was "in-phase". If the orientation was correct, the source and received traces were exactly "in-line". An example of non-desirable output is presented in Figure 4. The source signal and the received one are not simultaneously upward.

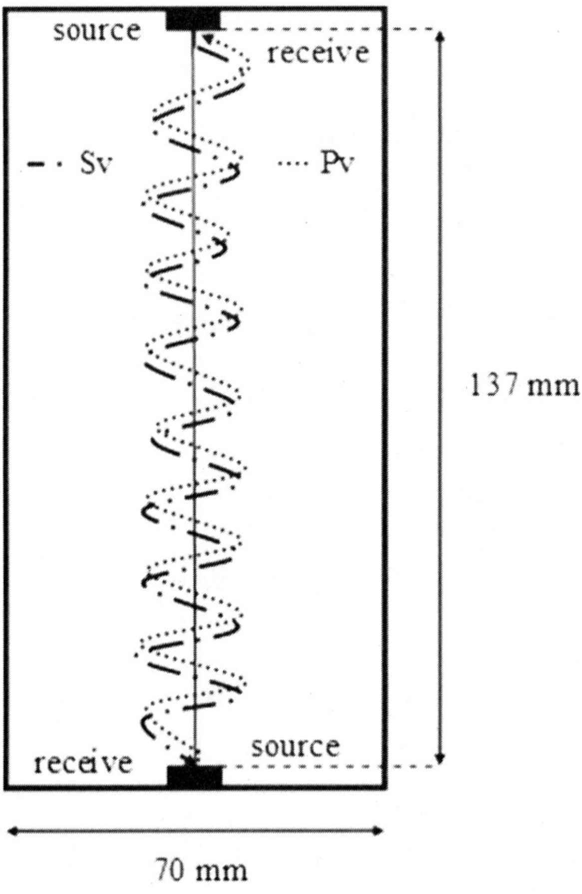

Figure 3. Frontal sample view with bender element (BE) arrangement and an example of the waves' path.

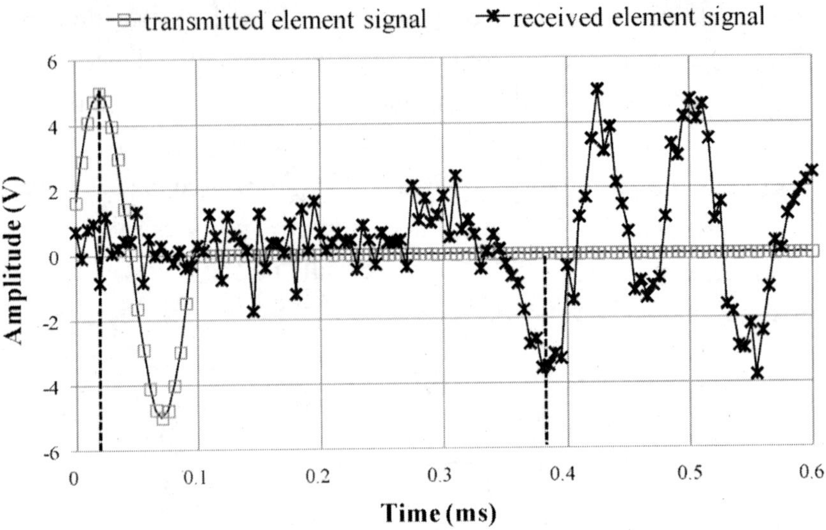

Figure 4. Non-desirable orientation of the top cap and base pedestal element.

2.2. Methods of Interpretation of BE Results

The BE technique has many advantages, which have been mentioned so far, but the interpretation of BE results can often be questionable [18,22,27]. A number of methods for the analysis of bender element results are commonly used in the time or frequency domain [17,28]. The time domain methods are classified as direct measurements, which use plots of electrical signals *versus* time [19,29]. These methods employ a time-based axis in order to identify the propagation time, hence [6]. In frequency domain approaches, on the other hand, the spectral breakdown of the signals is analyzed, and the shift in the phase angle between the trigger and response signal is calculated [19,30]. The time domain techniques are generally simpler and more straightforward, as the travel time can be directly defined by studying the time interval between characteristic points in the transmitted and received wave traces. However, the frequency domain methods appear to be more detailed, since they employ the support of signal processing and spectrum analysis tools. Additionally, they enable automated data acquisition and processing [18]. It is important to note that no method of interpreting the BE test results is yet proven superior to the others [6,19,31]. Reliable determination of travel time is very significant, due to BEs being installed in laboratory equipment and their application for studying relatively small specimens, which means that the travel distance is quite small. Below, the authors briefly present the methods of determining the arrival time of the shear wave that were employed in the paper. These are further illustrated in Figure 5.

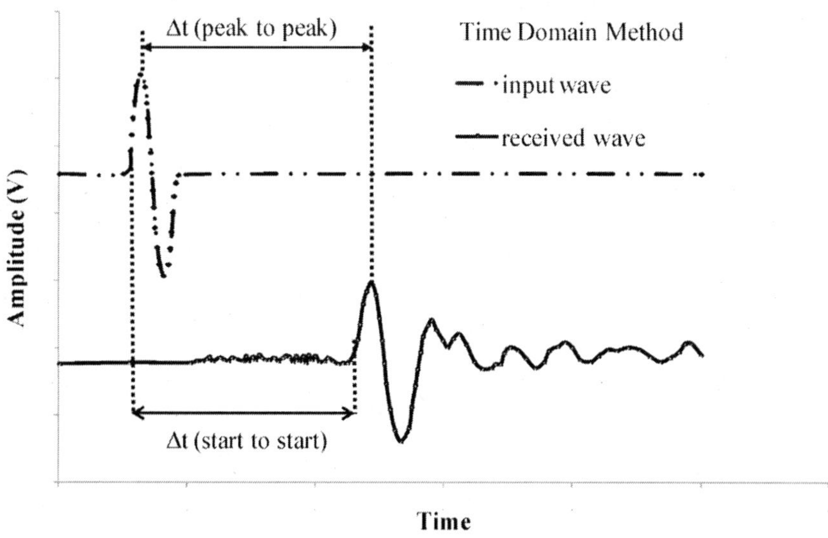

Figure 5. Typical identification techniques of travel time in the time domain method.

The peak-to-peak method [19,28] is based on the assumption that the received signal bears a high resemblance to the transmitted one. The travel time may be treated as the time elapsed between any two corresponding characteristic points in the signals. By doing this, the near field problems should be reduced. The characteristic points that are most commonly used to identify the wave travel time are the first positive peak in the input of S-wave signals and in the output [32]. In some circumstances, however, properly defining the first major peak becomes quite difficult, as the received signal has several consecutive peaks, which differ slightly in amplitude. The reliability of choosing the right first peak as the first major one can be influenced by sample geometry and size or, for example, by the energy-absorbing nature of the soil. The arriving signal is then distorted to various extents due to the increase of damping with distance. The quality of received signals significantly affects this technique [28].

The principle of the start-to-start method, also known as the visual picking method, is the acceptance of the moment, when the first major deflection of the received signal occurs, as the shear wave arrival time [19,33]. The presented method is characterized by simplicity, which is the main reason behind its popularity. Depending on the installation and polarity of the bender elements, the first important deviation from zero amplitude may be positive or negative [28]. In many cases, it is simply a single sine pulse. This represents the start of energy transfer from the source to the soil. However, in case of the output wave signal, it is the moment when the investigated receiver begins exhibiting motion. Said moment represents the instant of the energy transfer from the soil to the receiving BE [32].

During the bender element test, shear wave velocity is calculated from the simple measurement of propagation distance (Δs) and propagation time (Δt). Based on a number of previous works, it is generally accepted that the travel distance is the distance between the tips of two BEs [11]. There are various waveforms, for example sine and square waves, with various frequencies, that have been recommended as an excitation signal [2].

2.3. Characterization of Test Material

In order to investigate the elastic modulus of cohesive material, associated with the passage of seismic waves (an S-wave and a P-wave), silty clay (siCl) samples, from the test site located in the center of Warsaw (the capital of Poland), recognized by particle size analysis using the sieves and hydrometer methods (European Standard Eurocode 7, [34]), were used. The description of the research area can be found in [25]. The index properties of the examined soils are summarized in Table 1. In order to compute the fundamental indices (w_P, w_L), standard test methods were employed, namely plastic limit test of soil, Casagrande liquid limit test and fall cone liquid limit test. The grain size distribution of the soil specimens is shown inFigure 6. The proportional content of each fraction is as follows: Gr = 0%, Sa = 13%, Si = 66%, Cl = 21%. The soil used during the tests, which is of Quaternary origin, was sampled in an undisturbed state using a standard Shelby tube. All of the samples were acquired at a depth of around 6.0 m and selected cautiously with the consideration of the soil structure uniformity, their physical properties and their double phase. Tubes were pressed carefully and gently into the pre-drilled holes. Subsequently, the samples were sealed and stored in a humidity room until needed [35].

Table 1. Basic properties of the tested soils.

Parameter	Value
w (%)	17.52
w_P (%)	17.14
w_L (%)	33.00
I_P (%)	15.86
I_L (%)	0.02
I_C (%)	0.98
ρ (kg/m^3)	2140

w is the water content; w_P is the plastic limit; w_L is the liquid limit; I_P is the plastic index; I_L is the liquidity index; I_C is the consistency index; and ρ is the mass density.

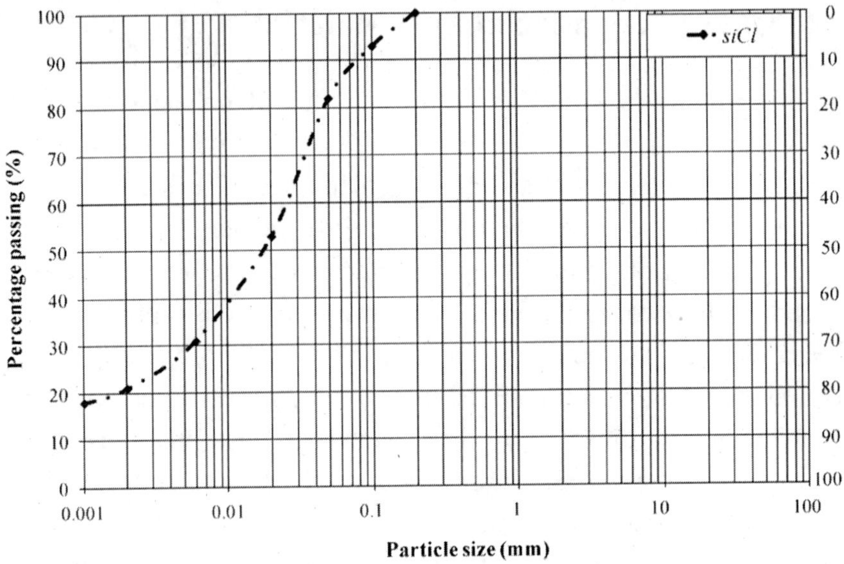

Figure 6. Grain size distribution of the tested soils.

2.4. Test Setup and Procedure

The test procedure followed that of a drained isotropic resonant column test. An upright cylindrical specimen of soil, with an aspect ratio of 2:1 (length:diameter), was employed. The proportions between length and diameter were as follows: 140 mm × 70 mm. Undisturbed material was set up in the resonant column cell, then saturated by the back pressure method in order to achieve the level of full saturation and subsequently consolidated to predetermined isotropic stress levels of 30, 120, 180, 240, 360 and 410 kPa. Back pressure was increased slowly, in order to ensure the proper saturation of the sample, until Skempton's *B* value reached 0.84. This represents, with respect to Head [36], the saturation level being equal to approximately 97%. During the consolidation stage, the axial deformation and volume change of the sample were measured. The back pressure was kept at a constant level of 290 kPa. At the end of each consolidation stage, seismic wave velocities were checked, in the undrained conditions, as well as using the bender elements located at the top and bottom of the soil specimen. The BE transmitter was excited with a sine pulse, whose magnitude equaled 15 V. A change of voltage applied to the transmitter caused bending and transmission of a shear wave through the sample. The receiver, located at the other end of the specimen, registered the arrival of the shear wave as a change in voltage [37,38,39]. For each isotropic confining pressure level, a range of input frequencies between 100 and 1 kHz was tested, in order to identify the value of greater amplification in the received signals. This value should represent the clearest output signals. In the case of some input frequencies, particularly the higher ones, compression wave velocity was measured additionally. Due to the limitations of the software and hardware (S- and P-waves were transmitted by the same piezoelectric element) provided by the GDS company, the authors were unable to conduct the measurements in the same period (T) of S- and P-waves. Hence, there is an apparent limit in the predetermined period or frequency of propagating various waves through the soil sample.

In Figure 7, the oscilloscope data obtained during the discussed studies are shown. The input signals at different frequencies were sent under the same stress conditions. During the analysis of this figure, it appeared to the authors, that at higher frequencies, the determination of the arrival time was more precise. Changes in the input signal frequency did not produce any change of polarization in the shear waves. For all applied frequencies, the input and output signals indicated the same amount of polarization.

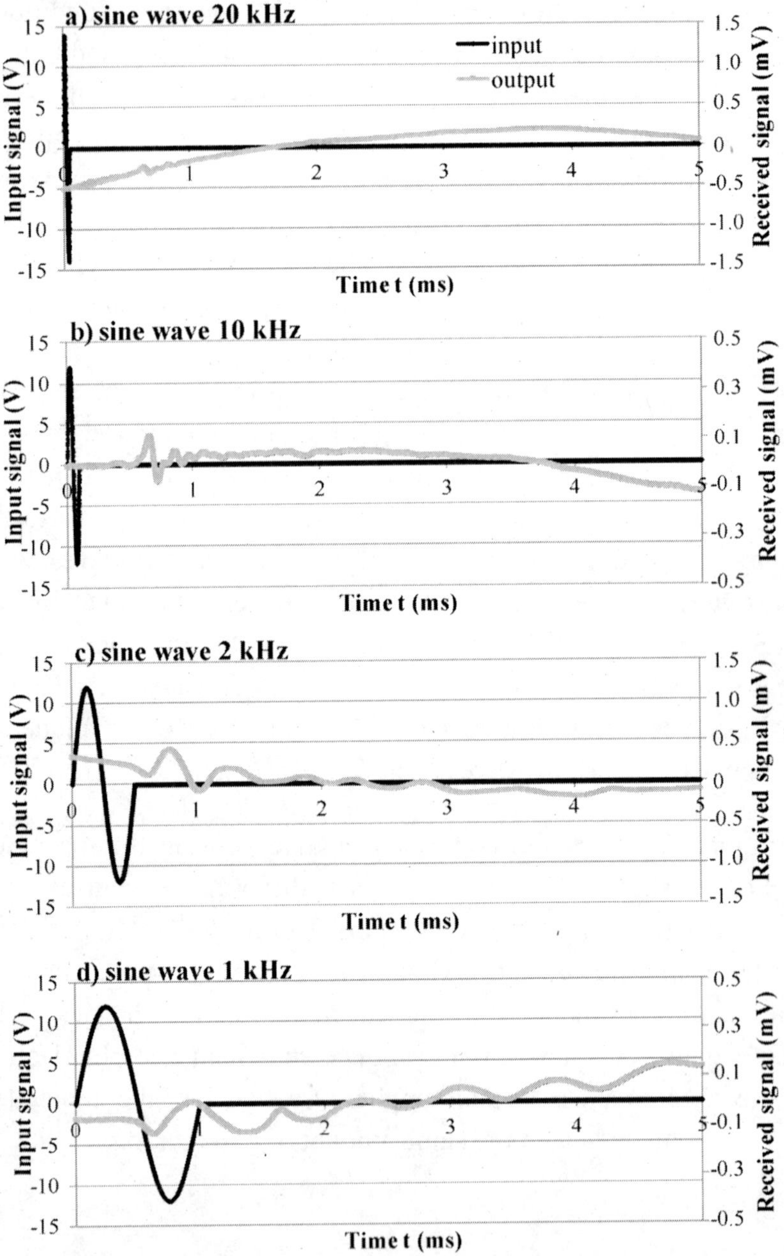

Figure 7. The oscilloscope signals at various frequencies of input signals: (**a**) 20 kHz, (**b**) 10 kHz, (**c**) 2 kHz, (**d**) 1 kHz and at the same effective stress value.

To interpret the BE results, the authors applied two commonly-known techniques used in time domain methods, which they described in the earlier subsection. For the start-to-start method, the noise level is very significant. Leong *et al.* [40] verified the criteria for the improvement of the interpretation of the BE test, among which there is the signal-to-noise ratio (SNR) criterion. According to their research, the SNR should be of at least 4 dB for the receiver signal. In the authors' research, the average SNR for the start-to-start method was equal to 10.44 dB. This level of the SNR allowed picking a very reliable and comfortable wave onset. Values of the wave velocity were calculated on the basis of the tip-to-tip distance between the transmitter and the receiver bender element [41]. The shear wave velocity was evaluated from the relationship presented below [11,42]:

$$\upsilon_S = \frac{h}{t} \tag{5}$$

where h is the distance between the transmitter and the receiver and t is the travel time.

Finally, the unloading process was carried out. After each unloading stage, BE tests were also performed, followed by the measurement of travel times, which were determined using peak-to-peak and start-to-start interpretation procedures.

BE tests were conducted to provide information on the shear (G) and Young's (E) moduli. From the S-wave velocity (υ_S), the small strain shear modulus (G_{max}), was determined, using the elastic wave velocity according to the equation [43,44]:

$$G_{max} = \rho \cdot \upsilon_S^2 \tag{6}$$

where ρ is the soil mass density.

By transforming Equation (3) and combining it with Equation (5), the authors estimated the value of the small strain Young's modulus (E_{max}) as follows [45]:

$$E_{max} = 2 \cdot \rho \cdot \upsilon_S^2 \cdot (1 - \nu) \tag{7}$$

where ν is Poisson's ratio.

According to a well-known relation [4], Poisson's ratio is equal to:

$$\nu = \frac{\left(\frac{\upsilon_P}{\upsilon_S}\right)^2 - 2}{2\left[\left(\frac{\upsilon_P}{\upsilon_S}\right)^2 - 1\right]} \tag{8}$$

3. RESULTS AND DISCUSSION

3.1. Frequency Dependency

In Figure 8 and Figure 9, the variation of the measured shear wave velocity (υ_S) with a predetermined period of propagating wave at different specified stress levels is shown. In Figure 10, however, the equivalent plot for measured compression wave velocity (υ_P) is illustrated. The presented results are related to both procedures adopted for determining the travel time.

Inspection of Figure 8 and Figure 9 indicates that, regardless of the method used to measure the travel time, the shear wave velocities decrease with the predetermined period (*T*), which transfers to frequency (*f*). Oscillation frequency is in fact the inverse of period. The linear approximation fits accurately most of the presented data obtained using the peak-to-peak and start-to-start methods. The coefficient of determination (R^2) reaches high values, from 70%

upwards, which shows that for the studied cohesive material, the linear regression functions explain the majority of υ_S variation. In general, the differences in the shear wave velocities with different frequencies are not large, from 18 to 27 $m \cdot s^{-1}$ and from 32 to 38 $m \cdot s^{-1}$, depending on the technique used to measure the travel time.

In said graphs, the influence of stress level (p') is visible, as well. The average effective stress level significantly affects υ_S. For the highest pre-set pressure, *i.e.*, 410 kPa, the authors received the highest values of shear wave velocity, with average values around 307 $m \cdot s^{-1}$ (peak-to-peak method) and 319 $m \cdot s^{-1}$ (start-to-start method). The lowest values of v_S were recorded for the smallest given stress (30 or 180 kPa), their average being around 160 $m \cdot s^{-1}$ (peak-to-peak method) and 246 $m \cdot s^{-1}$(start-to-start method). The difference in υ_S decreases with the increasing pressure value.

The results of the current study show also that for the period of propagation lasting up to 0.1 ms (*i.e.*, f = 10 kHz), both selected procedures used in the time domain methods consistently produce the values of υ_S, which are less dependent on frequency than in the case of a larger period of wave (Figure 8 and Figure 9 on the right). The frame in these figures indicates the highest values of shear wave velocity for each mean effective stress level. The authors suggest in this way certain frequencies at which the bender elements tests should be performed. Hence, the recommended frequency values, for which the highest shear wave velocities can be obtained, are in the range of 20 to 50 kHz. Certainly, it should be remembered that the optimum frequency of excitation depends on many factors, for example on the pressure applied in the cell during tests [46]. There is no rule at all for choosing the best wave [47]. In the literature, some authors have reported their experience with choosing a specific frequency. Brignoli *et al.* in 1996 [38] noted that when bender elements are measuring the samples, whose height is between 100 and 140 mm, the most interpretable output waves occur in the range from 3 to 10 kHz. Jovočič *et al.* [33], however, observed that with a higher frequency, the near field effect is less discernible. For stiffer materials, the required frequency must be higher than the reported 10 kHz [33], although the problem of overshooting becomes more important. In addition to this, a high frequency wave attenuates faster than the low frequency one, provided the medium is the same. The fact that the use of a high frequency gives the clearest output signals remains undisputed.

In the case of compression waves (Figure 10), the authors carried out research in the frequency scope of up to 10 kHz. A very large dispersion of the results and a lack of dependence of υ_P on the frequency ratio, as well as on mean effective stress can be clearly noticed. The authors could not find a proper relationship of υ_P *vs.* frequency and mean effective stress. Empirical equations expressing the variation of P-wave velocity with frequency are characterized, however, by the low coefficients of determination (R^2), listed in Figure 10. In general, the differences in compression wave velocities with the different frequencies reach high values, from 54 to 110 $m \cdot s^{-1}$ and from 19 to 57 $m \cdot s^{-1}$, depending on the technique used to measure the travel time.

In Figure 11 and Figure 12, the effect of excitation frequency on shear modulus (Figure 11) and Young's modulus (Figure 12) is presented. The excitation frequencies varied from 1 to 100 kHz. The specimens were subjected to various isotropic effective confining pressures. It was observed (black frame in the figures) that the best signal, and consequently the greatest values of shear modulus (G), were received at the excitation frequency, whose values were equal 20 kHz for the peak-to-peak method (Figure 11a) and 50 kHz for the start-to-start method (Figure 11b). The excitation frequency being equal to (Figure 12b) or smaller than 50 kHz (Figure 12a) gave the greatest values of Young's modulus (E), as well. It is worth paying attention to a decrease in the modulus values along with the decreasing frequency, although G_{max} and E_{max} values were not obtained for the highest frequency, as already mentioned above. The average G_{max} value obtained from all applied values of pressure ranged from 139 MPa to 194 MPa, depending on the measurement of the travel time, while average E_{max} was around 405 MPa and 557 MPa. In the case of the minimal dynamic properties of the studied soils, using the lowest frequencies resulted in the smallest values of *G* and *E*, with the average minimal *G* ranging from 116 MPa to 152 MPa and the minimal *E* ranging from 379 MPa to 509 MPa.

It should be also pointed out that the values of *E* are arranged in some trend with increasing frequency, whereas the values of υ_P do not show any dependence on frequency. This may seem incorrect, because Young's modulus is relevant to the compression wave velocity, due to the motion of the particles. However, it must be remembered that the authors used the results of υ_P only indirectly, in order to determine the proper value of ν.

3.2. The Effect of Stress Level

In Table 2, the authors compiled the range of received S-wave and P-wave velocities after each consolidation stage. On the basis of Table 2, it can be concluded that the values of shear wave velocity obviously increase with applied pressure. At the same time, it is visible that the differences between the shear wave velocity values and each *p*' amount to 20 to 30 $m \cdot s^{-1}$. The values of compression wave velocity confirmed the earlier observation about the lack of dependence between υ_P and stress level. The typical P-wave velocity of the tested material reached the value of 1400 to 1500 $m \cdot s^{-1}$ for the peak-to-peak method and 1600 to 1700 $m \cdot s^{-1}$ for the start-to-start method.

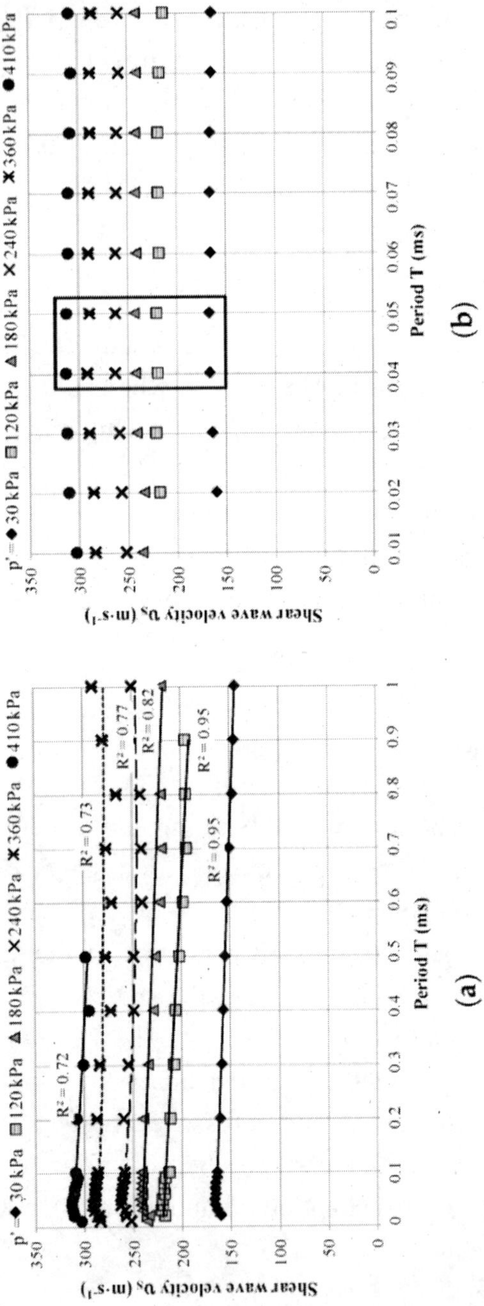

Figure 8. Variation of measured v_S with period: peak-to-peak method. (**a**) Period's range from 0 to 1 ms; (**b**) Period's range from 0.01 to 0.1 ms.

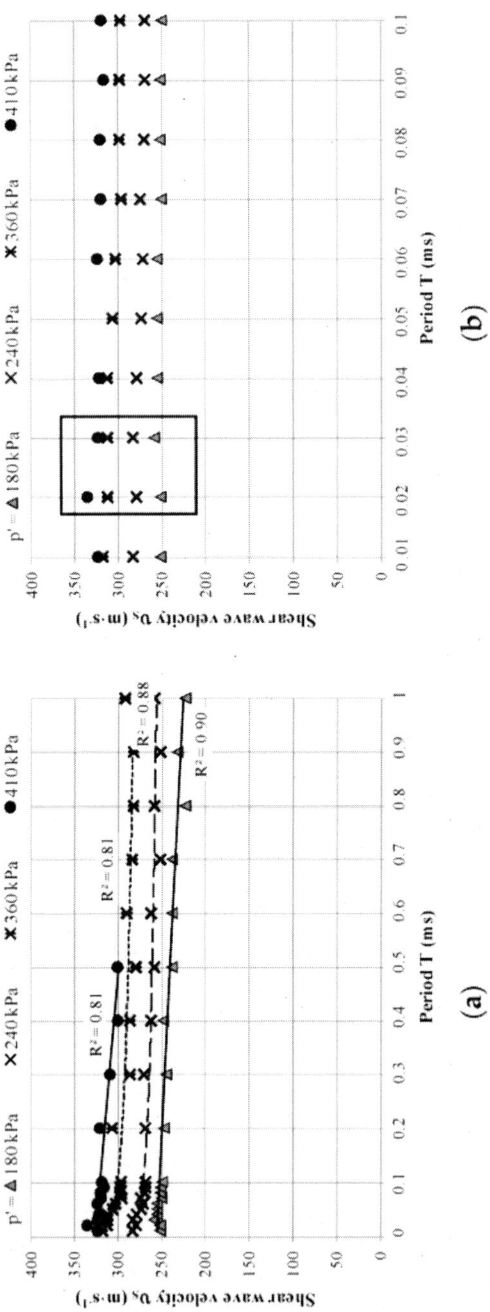

Figure 9. Variation of measured υ_S with period: start-to-start method. (**a**) Period's range from 0 to 1 ms; (**b**) Period's range from 0.01 to 0.1 ms.

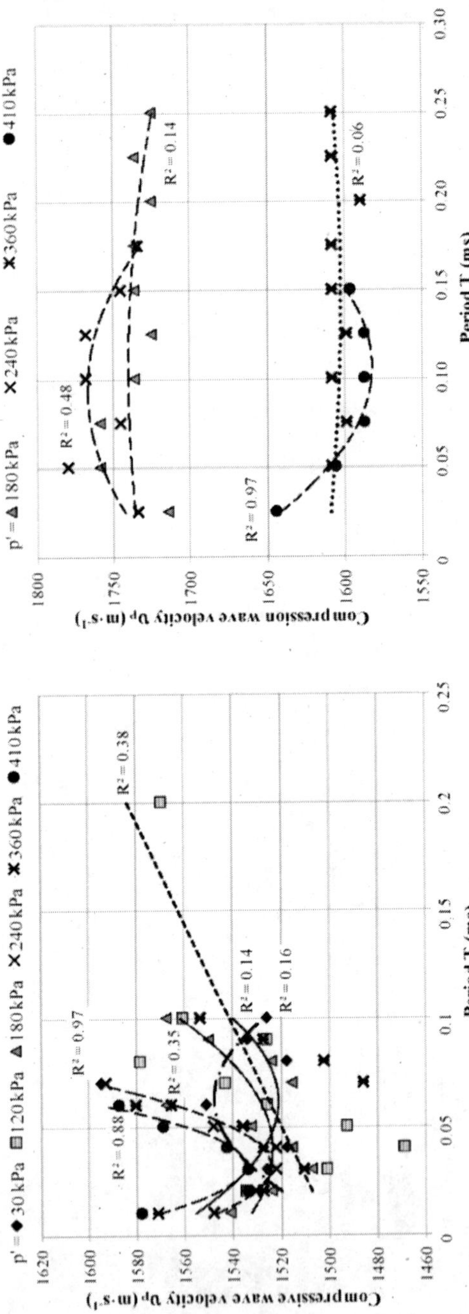

Figure 10. Variation of measured υ_P with period. (**a**) Peak-to-peak method; (**b**) start-to-start method.

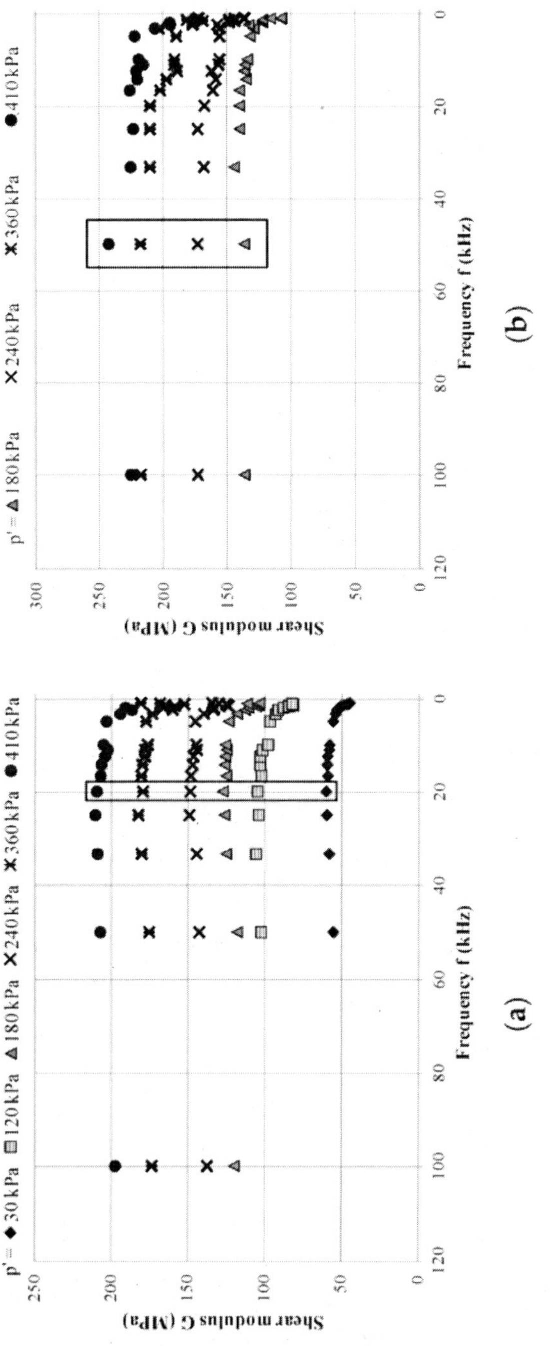

Figure 11. Variation of dynamic stiffness with frequency. (**a**) Peak-to-peak method; (**b**) start-to-start method.

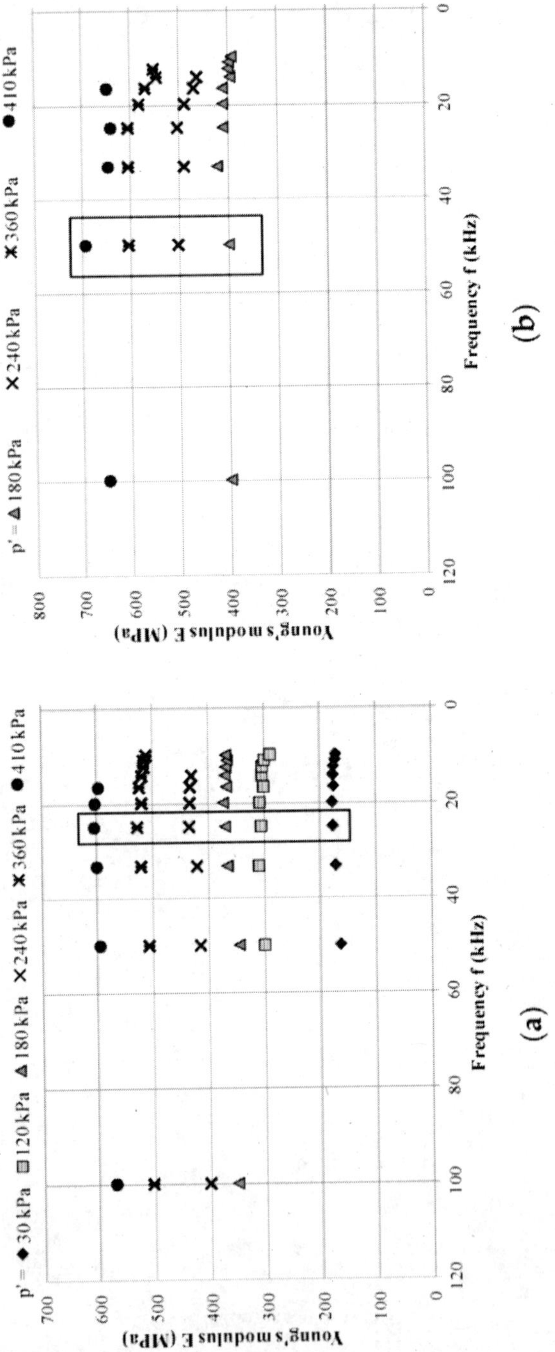

Figure 12. Variation of Young's modulus with frequency. (**a**) Peak-to-peak method; (**b**) start-to-start method.

Table 2. Seismic waves velocities for various mean effective stress levels. (**a**) Peak-to-peak method; (**b**) start-to-start method.

(a)

Mean Effective Stress	S-Wave Velocity	P-Wave Velocity	Average Ratio of Seismic Waves Velocities
p' (kPa)	v_S (m·s^{-1})	v_P (m·s^{-1})	v_P/v_S (-)
30	145–168	1518–1595	9.3
120	195–222	1469–1579	7.0
180	221–224	1508–1568	6.4
240	240–264	1522–1593	5.9
360	267–292	1486–1581	5.3
410	296–314	1435–1588	5.0

(b)

Mean Effective Stress	S-Wave Velocity	P-Wave Velocity	Average Ratio of Seismic Waves Velocities
p' (kPa)	v_S (m·s^{-1})	v_P (m·s^{-1})	v_P/v_S (-)
180	224–260	1715–1759	6.8
240	252–284	1734–1779	6.3
360	281–319	1590–1609	4.7
410	301–336	1588–1645	4.9

In Table 2, additionally, the average ratio list of seismic wave's velocities is enclosed. It is evident that the values of υ_P are several times higher, when compared with υ_S. The υ_P/υ_S velocity ratio has the variation interval ranging from 5.0 to 9.3. The authors obtained larger values of P-wave velocity than is observed in certain positions in the literature. Das and Ramana [48], for example, reported the characteristic compression wave velocity as ranging from 1220 $m \cdot s^{-1}$ to 1370 $m \cdot s^{-1}$ for saturated clay, although the ratio υ_P/υ_S was in the range of 8.0 to 9.0, thus similar to those obtained by the authors. Gary Mavko [49], on the other hand, in his study on Parameters That Influence Seismic Velocity mentions P-wave velocity for saturated clays as being in the range of 1100 to 2500 $m \cdot s^{-1}$. Campanella and Steward [50] detailed typical speeds of longitudinal and transversal waves to a depth of 40 m. According to their research, compressive wave velocity for saturated soils may have values between 1500 and 1900 $m \cdot s^{-1}$.

In Table 3, the results of the average dynamic Poisson's ratio computed for each applied mean effective stress is shown. ν_d was determined on the basis of the relationship between S-wave and P-wave velocities (Equation (8)). The values of ν_d seem to be consistent with the literature [4,51,52]. The authors described the values of this parameter in another publication [23], in which the results obtained for saturated clay were at the level of 0.40 to 0.50, which is similar to the value obtained during the current studies. It is interesting to note that there is no significant change in ν_d regarding the pressure. Variation in Poisson's ratio amounts to approximately 4% and approaches 0.5. Moreover, a number of different techniques for interpreting the BE results causes no definite difference in the values of Poisson's ratio. Based on the experiments conducted, a general trend can be observed: as far as tested clayey material is concerned, Poisson's ratios represent little appreciable change with regard to shear modulus (Figure 13). The values of ν_d were used afterwards in order to determine the dynamic modulus of elasticity (E).

Table 3. Dynamic Poisson's ratio for various mean effective stress levels. (**a**) Peak-to-peak method; (**b**) start-to-start method.

(a)

Mean Effective Stress	Average Dynamic Poisson's Ratio
p' (kPa)	ν_d (-)
30	0.48
120	0.47
180	0.46
240	0.46
360	0.44
410	0.44

(b)

Mean Effective Stress	Average Dynamic Poisson's Ratio
p' (kPa)	ν_d (-)
180	0.47
240	0.46
360	0.44
410	0.44

The effect of mean effective stress on the small strain shear modulus and on the small strain modulus of elasticity of the examined silty clays is displayed in Figure 14 and Figure 15. The presented data were obtained for one excitation frequency, selected on the basis of its greatest values of dynamic properties regarding the analyzed soil specimens, as described in the previous section. BE measurements show that G_{max} and E_{max} values increase with confining pressure as a linear function. Shear modulus at small strains, which was calculated for the tested material using Equation (6), does not exceed the interval of 60 to 242 MPa, regarding the travel time measurement. The modulus of elasticity at small strains, on the other hand, varies from 178 MPa to 694 MPa.

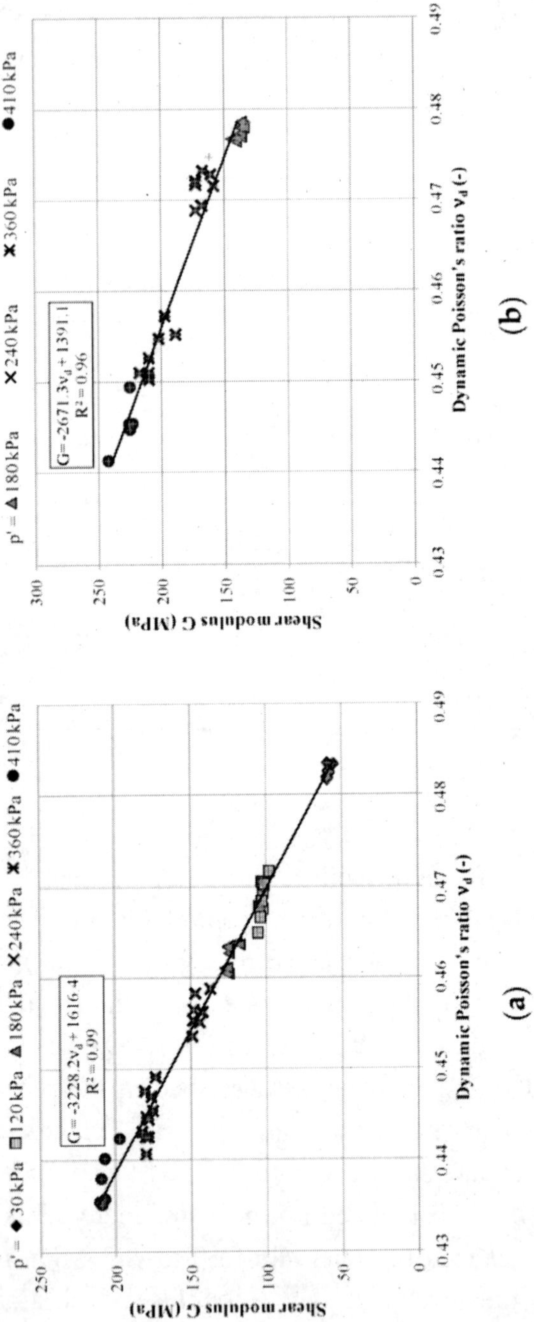

Figure 13. Shear modulus *versus* dynamic Poisson's ratio. (**a**) Peak-to-peak method; (**b**) start-to-start method.

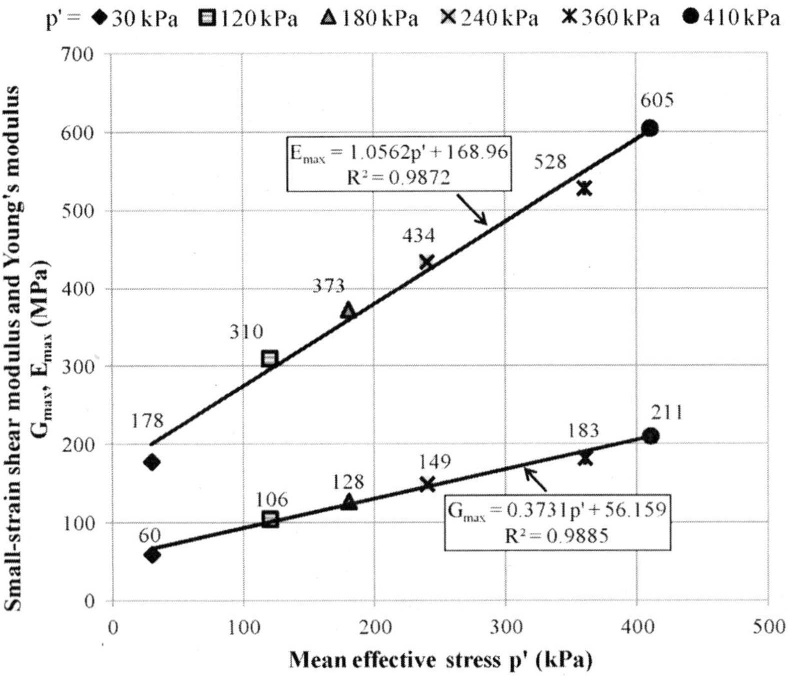

Figure 14. Elastic moduli from the peak-to-peak method under different stress levels.

3.3. Comparison of Results from Different Methods

The input signals employed in all of the tests were single sinusoidal pulses with different frequencies, obtained using peak-to-peak and start-to-start methods. The amplitude of input signals in both cases was fixed at ±15 V. The tip-to-tip distances between the transducers were used to calculate the shear wave velocity of the samples (Equation (5)). All of the test traces were examined in the time domain, in order to obtain the travel time of the wave through the specimens.

In Figure 16, S-wave velocities determined by means of the two analyzed time domain techniques are portrayed. It is clearly visible that travel time derived from the start-to-start method must be longer due to the presence of the greatest values of the shear wave velocities. The difference in the obtained values of υ_S between these two methods remains at the average level of 21.25 m·s^{-1}, in favor of the start-to-start method (Figure 17a). Naturally, these spotted differences vary, depending on the mean effective stress.

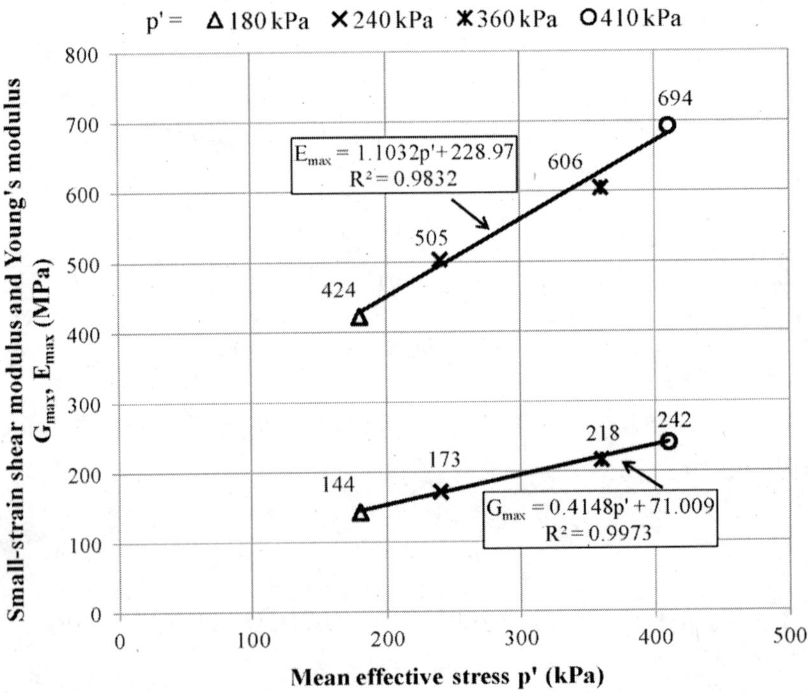

Figure 15. Elastic moduli from the start-to-start method under different stress levels.

In Figure 17b, the evolution of the shear stiffness modulus based on travel time deduced by means of the peak-to-peak and start-to-start techniques is summarized. In this comparison, the second of the aforementioned methods comes off as the better one. The average difference in stiffness of the examined cohesive material falls in the range of about 33.44 MPa. For the E_{max} (Figure 17c), the discrepancy between the results of two different interpretation methods achieves a maximum percentage of about 16% (89.2 MPa).

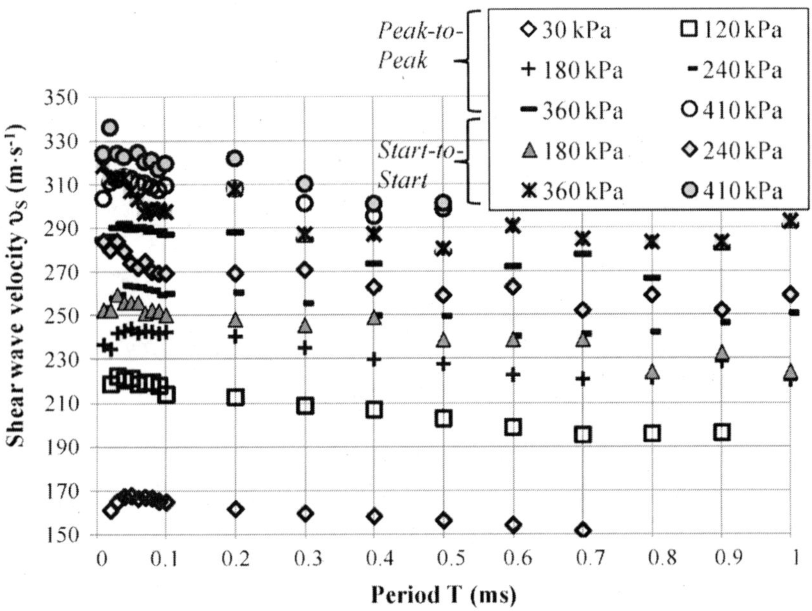

Figure 16. Comparison of υ_S obtained using various methods of travel time determination.

In conclusion, it is worth noting that in the case of the start-to-start method, the findings of υ_S and G_{max} lie on a straight line, but of a different gradient. This observation is not reflected in the outcomes regarding E_{max}. The results obtained with the use of the peak-to-peak method appear to be more scattered. On the basis of Figure 17, it can be noted that the S-S method (start-to-start method) of identification results in a relatively smaller variation, when compared to the P-P technique (peak-to-peak method). The latter method yields slightly smaller values of the dynamic parameters.

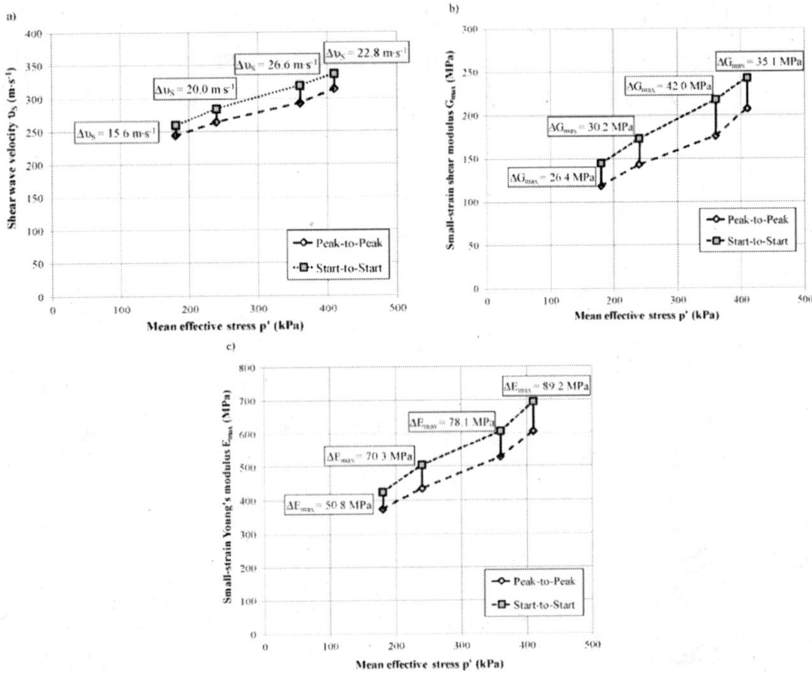

Figure 17. Comparison of results obtained by means of the peak-to-peak and start-to-start methods. (**a**) Shear wave velocity; (**b**) small strain shear modulus; (**c**) small strain Young's modulus.

3.4. The Effect of Unloading

The aim of the current study, besides finding S-wave velocity and the small strain shear modulus when loading soil samples, was also to find υ_S and G_{max} while unloading. The unloading process of the tested material was carried out to the pressure level of 120 kPa. In Figure 18 and Figure 19, the values of computed parameters from the loading and the subsequent unloading stages are presented.

During the inspection of Figure 18 and Figure 19, the authors noticed that $\upsilon_{S,\ unloading}$ and consequently $G_{max,\ unloading}$ carry greater values than $\upsilon_{S,\ loading}$ and $G_{max,\ loading}$. The variance between silty clay dynamic stiffness during loading and unloading stages falls within the range between 16.8 and 22.8 MPa. The biggest difference was obtained for the mean effective stress equal to 180 kPa, the smallest one for p' = 360 kPa. The unloading process causes the increase of the results from 8% to 16%. Thus, it should be highlighted that the unloading of soil considerably improves its stiffness.

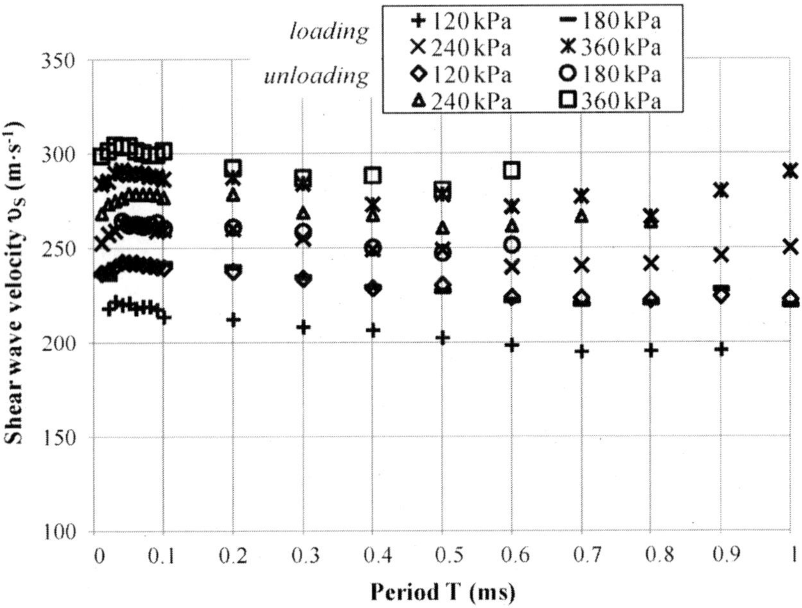

Figure 18. Shear wave velocity *versus* period at the loading and unloading stage.

4. CONCLUSIONS

In this study, the characterization of the dynamic properties of cohesive material, namely silty clays, was inspected. The conducted experiments included the dynamic tests in the GDS Resonant Column Apparatus equipped with bender elements. The authors aimed to verify the dynamic behavior of clayey soil in connection with wave propagation produced by piezoelectric transducers. Shear wave velocity is related to the shear modulus of soil. Consequently, the measurements of the S-wave velocity provided a convenient method for determining soil stiffness. Additionally, the calculation of the P-wave velocity and, subsequently, the dynamic Poisson's ratio allowed for the examination of Young's modulus with regard to the tested cohesive material.

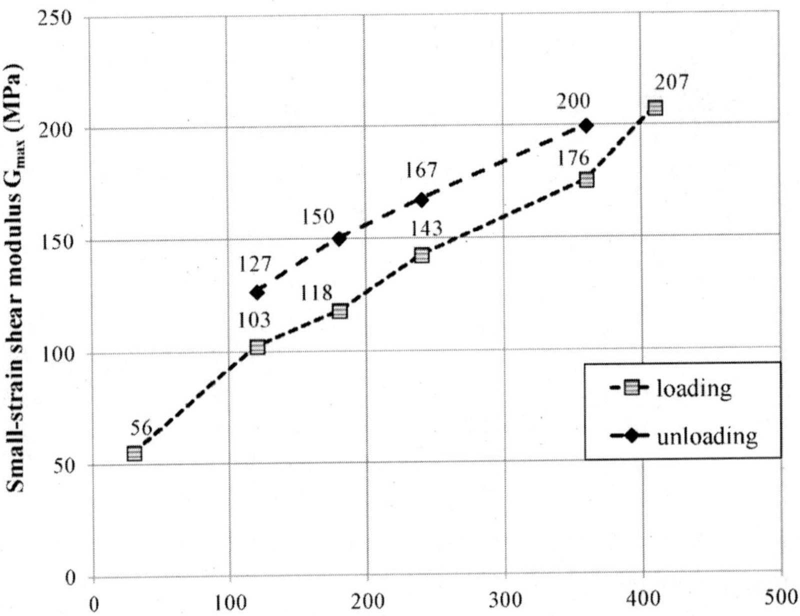

Figure 19. Small strain shear modulus *versus* mean effective stress at the loading and unloading stage.

The findings of this research, in line with the literature in relation to the methods of interpretation of bender element test results, indicate that, for tests with piezoelectric transducers installed in the resonant column apparatus on silty clays, both analyzed techniques used as a part of the time domain methods, *i.e.*, peak-to-peak and start-to-start, yield similar results. They are also considered as the most comprehensive and appropriate methods to identify the travel time of the shear waves propagating through the sample. The authors, however, did not inspect any other methods, like the cross-correlation or π-point phase, but the velocities established by those methods are comparable to the findings from the resonant column measurements [25]. The dynamic parameters obtained by the start-to-start method were always slightly larger (up to about 20%) than those obtained using the peak-to-peak method. According to the authors, this is a consequence of the broadening of the excitation pulses after they are propagated through the sample. The lowest wavelength in the presented experiments was about 1.0 mm, which was at least 10-times bigger than the largest grain size of soil sample (that was less than or equal to 100 μm). In addition, no inhomogeneous pockets were found in the sample constitution after a destructive analysis of the material. In these conditions the macroscopic sample was expected to behave as a conducting media, and no enhanced scattering was expected at the grain boundaries. Since the studied sample was homogeneous, in terms of the wavelength, the difference between the results from the two discussed techniques can be explained by taking into account different wave velocities and different attenuations of the frequency components that constitute the excitation peaks. Although each excitation pulse used in the experiments was considered to have a well-defined frequency, in reality, the pulses are composed of a continuity of different frequencies with different amplitudes. This happens due to the finite size of the excitation pulse in the time domain, which adds additional frequencies to it. In the start-to-start method, the results were strongly influenced by the fastest travelling wave, while in the peak-to-peak method, the frequency component with the highest amplitude was the one considered. From the Figure 9 and Figure 10, it can be clearly noticed that the wave velocity from the S-S method had an almost flat dependency as a function of frequency; however, in the P-P method, it has a change of about 20% in the frequency range experimented. This suggests that, in the S-S method, the frequency component responsible for the excitation of the detector is independent of the frequency chosen for the excitation pulse (due to the unintentional frequencies present in the excitation pulse). In certain experimental circumstances, such as during the measurements of G, the peak-to-peak method led to more scattered results, which were probably connected to the interpretation of the findings made by the authors. It must be always kept in mind, that even when using the same identification method, reading points differed with the laboratory [22]. The scatter in the S-S method was definitely smaller in comparison to another method. Presented results show and confirm observations made by previous researchers [6,22], acknowledging the status of the start-to-start method as the most prevalent and consistent method for the determination of travel time.

With reference to the dependence of the dynamic behavior of clayey material on various factors, the following conclusions can be drawn:

- The data show that the input signal frequency, in particular frequencies, which are below the level of $T = 0.1$ ms, does not significantly affect the stiffness values obtained by both adopted identification methods. The decrease in frequency results in greater dependence on frequency.

- The best measurements of travel time for the tested soils were received at the frequencies of the input signal in the range of 50 to 20 kHz. In most of the considered cases, the greatest values of the dynamic parameters were obtained when the frequencies from this range were applied.

- Mean effective stress has a visible influence on the seismic wave velocities and on the elastic modulus of the examined soils.

- There are linear relationships between the mean effective stress and the small strain shear modulus and the small strain modulus of elasticity.

- The compression waves propagate through the analyzed clayey material with far larger velocities than in the case of shear waves; at certain pressure levels, these velocities can be even up to 10-times larger.

- The dynamic Poisson's ratio for the tested cohesive soils is almost constant, being approximately equal to 0.46.

- The dynamic Poisson's ratio linearly increases with the decrease of the shear modulus.

- Concerning the unloading process, it profitably affects the stiffness of silty clays. The dynamic properties rise in their values, from 8% to 16%, owing to unloading.

Based on the results as described above, the standard procedure for the BE tests was proposed here in order to obtain appropriate test results in the present conditions. The authors discussed the important issue regarding the applied sciences, precisely the seismic research and dynamic response of porous material due to dynamic loads, such as earthquakes, traffic loads or machine vibrations. They attempted to characterize the dynamic properties of a fine-grained material, whose average porosity is approximately 0.44. This differs from the commonly-encountered in the literature dynamic characteristics of granular materials, non-cohesive soils, with a larger porosity value, described for example in [53,54]. The authors used generally-applied laboratory tests in order to investigate a still new material, as well as to expand the knowledge of the interpretation of this new material's results.

The acquired knowledge of the behavior of the tested material should be verified in the process of modeling, using an appropriate model, such as the one proposed in [55].

ACKNOWLEDGMENTS

The discussion about the results obtained was carried out with the valuable support of Bruno Cury Camargo (LNCMI—Laboratoire National des Champs Magnetiques Intenses, Toulouse, France).

AUTHOR CONTRIBUTIONS

Wojciech Sas was the inventor of the research presented in this paper. Katarzyna Gabryś and Emil Soból conducted the described research and also performed geotechnical laboratory studies. Wojciech Sas and Katarzyna Gabryś prepared the manuscript. Katarzyna Gabryś and Emil Soból performed the analysis of the tests results. Alojzy Szymański consulted the research program and proofread the manuscript.

CONFLICTS OF INTEREST

The authors declare no conflict of interest.

REFERENCES

1. Salacak, M.; Pinar, S.B.S. Soil parameters which can be determined with seismic velocities. *Jeofizik* **2012**, *16*, 17–29.

2. Youn, J.-U.; Choo, Y.-W.; Kim, D.-S. Measurement of small-strain shear modulus G_{max} of dry and saturated sands by bender element, resonant column, and torsional shear tests. *Can. Geotech. J.* **2008**, *45*, 1426–1438.

3. Burland, J.B.; Longworth, T.I.; Moore, J.F.A. Study of ground and progressive failure caused by deep excavation in Oxford clay. *Géotechnique* **1977**, *27*, 557–591.

4. Oshaki, Y.; Iwasaki, R. On dynamic shear moduli and Poisson's ratio of soil deposits. *Soils Found.* **1973**, *13*, 61–73.

5. Sas, W.; Głuchowski, A.; Radziemska, M.; Dzięcioł, J.; Szymański, A. Environmental and Geotechnical Assessment of the Steel Slags as a Material for Road Structure. *Materials* **2015**, *8*, 4857–4875.

6. Yamashita, S.; Kawaguchi, T.; Nakata, Y.; Mikami, T.; Fujiwara, T.; Shibuya, S. Interpretation of international parallel test on the measurement of G_{max} using bender elements. *Soils Found.* **2009**, *49*, 631–650.

7. Benz, T. *Small-Strain Stiffness of Soils and Its Numerical Consequences*; Universität Stuttgart: Stuttgart, Germany, 2007.

8. Menzies, B. Near-surface site characterization by ground stiffness profiling using surface wave geophysics. In *H. C. Verma Commemorative Volume*; Indian Geotechnical Society: New Delhi, Indian, 2000; pp. 1–14.

9. Leong, E.C.; Cahyadi, J.; Rahardjo, H. Measuring shear and compression wave velocities of soil using bender-extender elements. *Can. Geotech. J.* **2009**, *46*, 792–812.

10. Ha Giang, P.H.; van Impe, P.; van Impe, W.F.; Menge, P.; Haegeman, W. Effects of grain size distribution on the initial small strain shear modulus of calcareous sand. In Proceedings of the 16th ECSMGE Geotechnical Engineering for Infrastructure and Development, Edinburg, UK, 13–17 September 2015; pp. 3177–3182.

11. Sas, W.; Gabryś, K.; Szymański, A. Laboratoryjne badanie sztywności gruntu według Eurokodu 7 [In Polish] Laboratory tests of soil stiffness by Eurocode 7. *Acta Sci. Pol. Ser. Archit.* **2013**, *12*, 39–50.

12. Sas, W.; Gabryś, K.; Szymański, A. Comparison of resonant column and Bender elements tests on selected cohesive soil from Warsaw. *Electron. J. Pol. Agric. Univ. EJPAU* **2014**, *17*, #07. Available online: http://www.ejpau.media.pl/volume17/issue3/art-07.html (accessed on 23 July 2014).

13. Lawrence, F.V. *Propagation of Ultrasonic Waves through Sand*; Research Report R63-08; Massachusetts Institute of Technology: Cambridge, MA, USA, 1963.

14. Lawrence, F.V. *Ultrasonic Shear Wave Velocity in Sand and Clay*; Research Report R65-05; Massachusetts Institute of Technology: Cambridge, MA, USA, 1965.

15. Shirley, D.J. An improved shear wave transducer. *J. Acoust. Soc. Am.* **1978**, *63*, 1643–1645.

16. Dyvik, R.; Madshus, C. *Lab Measurement of G_{max} Using Bender Elements*; Norwegian Geotechnical Institute: Oslo, Norway, 1985; pp. 186–196.

17. Styler, M.A.; Howie, J.A. Continuous Monitoring of Bender Element Shear Wave Velocities During Triaxial Testing.*Geotech. Test. J.* **2014**, *37*, 218–230.

18. Ferreira, C.; Martins, J.P.; Correia, A.G. Determination of the small-strain stiffness of hard soils by means of bender elements and accelerometers. In *Proceedings of the 15th European Conference on Soil Mechanics and Geotechnical Engineering*; Athens, Greece, 12–15 September 2013, Andreas, A., Michael, P., Christos, T., Eds.; IOS Press: Amsterdam, The Netherlands, 2011; Volume 1, pp. 179–184.

19. Viggiani, G.; Atkinson, J.H. Interpretation of bender element tests. *Géotechnique* **1995**, *45*, 149–154.

20. Lee, J.S.; Santamarina, J.C. Bender elements: Performance and signal interpretation. *J. Geotech. Geoenviron. Eng. ASCE***2005**, *131*, 1063–1070.

21. Leong, E.C.; Yeo, S.H.; Rahardjo, H. Measuring shear wave velocity using bender elements. *Geotech. Test. J.* **2005**, *28*, 488–498.

22. Camacho-Tauta, J.F.; Cascante, G.; da Fonseca, A.V.; Santos, J.A. Time and frequency domain evaluation of bender element systems. *Géotechnique* **2015**, *65*, 548–562.

23. Sas, W.; Gabryś, K.; Szymański, A. Determination of Poisson's ratio by means of resonant column tests. *Electron. J. Pol. Agricu. Univ. EJPAU* **2013**, *16*, #03. Available online: http://www.ejpau.media.pl/volume16/issue3/art-03.html(accessed on 20 August 2013).

24. Sas, W.; Gabryś, K. Laboratory measurement of shear stiffness in resonant column apparatus. *ACTA Sci. Pol. Ser. Arch.***2012**, *11*, 29–39.

25. Gabryś, K.; Sas, W.; Soból, E. Small-strain dynamic characterization of clayey soil from Warsaw. *ACTA Sci. Pol. Ser. Arch.* **2015**, *14*, 55–65.

26. GDS Instruments Ltd. Bender Elements. In *The GDS Bender Elements System Hardware Handbook for Vertical and Horizontal Elements*; GDS Instruments Ltd.: Hampshire, UK, 2005.

27. Cheng, Z.; Leong, E.C. A Hybrid Bender Element-Ultrasonic System for Measurement of Wave Velocity in Soils.*Geotech. Test. J.* **2014**, *37*, 377–388.

28. Chan, C.M. Bender Element Test in Soil Specimens: Identifying the Shear Wave Arrival Time. *EJGE* **2010**, *15*, 1263–1276.

29. Clayton, C.R.I.; Theron, M.; Best, A.I. The measurement of vertical shear-wave velocity using side-mounted bender elements in the triaxial apparatus. *Géotechnique* **2004**, *54*, 495–498.

30. Brocanelli, D.; Rinaldi, V. Measurement of low-strain material damping and wave velocity with bender elements in the frequency domain. *Can. Geotech. J.* **1998**, *35*, 1032–1040.

31. Arroyo, M.; Greening, P.; Wood, D.M. An estimate of uncertainty in current pulse test practice. *Riv. Ital. Geotec.* **2001**,*37*, 17–35.

32. Mitaritonna, G.; Amorosi, A.; Cotecchia, F. Multidirectional bender element measurements in the triaxial cell: Equipment set-up and signal interpretation. *Riv. Ital. Geotec.* **2010**, *1*, 50–69.

33. Jovičić, V.; Coop, M.R.; Simic, M. Objective criteria for determining G_{max} from bender element tests. *Géotechnique* **1996**,*46*, 357–362.

34. Polish Committee for Standardization. *Geotechnical Investigations—Soil Classification—Part 2: Classification Rules*; PN–EN ISO 14688-2:2004; Polish Committee for Standardization: Warsaw, Poland, 2004. (In Polish)

35. Sas, W.; Gabryś, K.; Szymański, A. Effect of time on dynamic shear modulus of selected cohesive soil of one section of Express Way No S2 in Warsaw. *Acta Geophys.* **2015**, *63*, 398–413.

36. Head, K.H. *Manual of Soil Laboratory Testing. Volume 3 Effective Stress Tests*; J. Wiley & Sons Ltd.: New York, NY, USA, 1998.

37. Viggiani, G. Panellist discussion: Recent advances in the interpretation of bender element tests. In *Pre-Failure Deformation of Geomaterials*; Satoru, S., Toshiyuki, M., Seiichi, M., Eds.; Balkema: Rotterdam, The Netherland, 1995; Volume 2, pp. 1099–1104.

38. Brignoli, E.; Gotti, M.; Stokoe, K.H., II. Measurement of shear waves in laboratory specimens by means of piezoelectric transducers. *ASTM Geotech. Test. J.* **1996**, *19*, 384–397.

39. Kawaguchi, T.; Mitachi, T.; Shibuya, S. Evaluation of shear wave travel time in laboratory bender element test. In Proceedings of 15th International Conference on Soil Mechanics and Geotechnical Engineering, Istambul, Turkey, 28–31 August 2001; Volume 1, pp. 155–158.

40. Leong, E.C.; Yeo, S.H.; Rahardjo, H. Measuring Shear Wave Velocity Using Bender Elements. *Geotech. Test. J.* **2003**.

41. Hasan, A.M.; Wheeler, S.J. Measuring travel time in bender/extender element tests. In Proceedings of the 16th ECSMGE Geotechnical Engineering for Infrastructure and Development, Edinburg, UK, 13–17 September 2015; pp. 3171–3176.

42. Choo, H.; Larrahondo, J.; Burns, S.E. Coating Effects of Nano-Sized Particles onto Sand Surfaces: Small Strain Stiffness and Contact Mode of Iron Oxide-Coated Sands. *J. Geotech. Geoenviron. Eng.* **2015**, *141*, 04014077:1–04014077:10.

43. Lee, C.-J.; Hung, W.-Y.; Tsai, C.-H.; Chen, T.; Tu, Y.; Huang, C.-C. Shear wave velocity measurements and soil-pile system identifications in dynamic centrifuge tests. *Bull. Earthq. Eng.* **2014**, *12*, 717–734.

44. Fedrizzi, F.; Raviolo, P.L.; Vigano, A. Resonant column and cyclic torsional shear experiments on soils of the Trentino valleys. In Proceedings of the 16th ECSMGE Geotechnical Engineering for Infrastructure and Development, Edinburg, UK, 13–17 September 2015; pp. 3437–3442.

45. Gabryś, K.; Szymański, A. Badania parametrów odkształceniowych gruntów spoistych w kolumnie rezonansowej [In Polish] Research of deformation parameters of cohesive soils in resonant column. *Inżynieria Morska i Geotech.* **2012**, *4*, 324–327.

46. Amat, A.S. Elastic Stiffness Moduli of Huston Sand. Master's Thesis, University of Bristol, Bristol, UK, 2007.

47. Ali, H.; Mahbaz, S.B.; Cascante, G.; Garbinsky, M. Low strain measurement of shear modulus with resonant column and bender element tests—Frequency effects. In Proceedings of the GeoMontreal 2013: Geoscience for Sustainability, Montreal, QC, Canada, 29 September–3 October 2013.

48. Das, B.M.; Ramana, G.V. *Principles of Soil Dynamics*, 2nd ed.; United States of America: Stamford, CT, USA, 2011.

49. Mavko, G. *Parameters that Influence Seismic Velocity. Conceptual Overview of Rock and Fluid Factors that Impact Seismic Velocity and Impedance*; Stanford Rock Physics Laboratory: Standford, CA, USA, 2009.

50. Campanella, R.G.; Stewart, W.P. Seismic cone analysis using digital signal processing for dynamic site characterization. *Can. Geotech. J.* **1992**, *29*, 477–486.

51. Poisson's ratio. Available online: http://www. Engineering toolbox.com (accessed on 21 September 2015).

52. Otsubo, M.; O'Sullivan, C.; Sim, W.W.; Ibraim, E. Quantitative assessment of the influence of surface roughness on soil stiffness. *Géotechnique* **2015**, *65*, 694–700.

53. Wichtmann, T.; Triantafyllidis, T. Stiffness and Damping of Clean Quartz Sand with Various Grain-Size Distribution Curves. *J. Geotech. Geoenviron. Eng.* **2014**, *140*, 06013003:1–06013003:4.

54. Qiu, T.; Huang, Y.; Guadalupe-Torres, Y.; Baxter, C.D.P.; Fox, P.J. Effective Soil Density for Small-Strain Shear Waves in Saturated Granular Materials. *J. Geotech. Geoenviron. Eng.* **2015**, *141*, 04015036:1–04015036:11.

55. Aggelis, D.G.; Tsinopoulos, S.V.; Polyzos, D. An iterative effective medium approximation (IEMA) for wave dispersion and attenuation predictions in particulate composites, suspensions and emulsions. *J. Acoust. Soc. Am.* 2004, *116*, 3443–3452.

CHAPTER 4

Unidirectional Wave Propagation in Low-Symmetric Colloidal Photonic-Crystal Heterostructures

Vassilios Yannopapas

Department of Physics, National Technical University of Athens, GR-15780 Athens, Greece; Tel.: +30-210-7721481

ABSTRACT

We show theoretically that photonic crystals consisting of colloidal spheres exhibit unidirectional wave propagation and one-way frequency band gaps without breaking time-reversal symmetry via, e.g., the application of an external magnetic field or the use of nonlinear materials. Namely, photonic crystals with low symmetry such as the monoclinic crystal type considered here as well as with unit cells formed by the heterostructure of different photonic crystals show significant unidirectional electromagnetic response. In particular, we show that the use of scatterers with low refractive-index contrast favors the formation of unidirectional frequency gaps which is the optimal route for achieving unidirectional wave propagation.

KEYWORDS

photonic crystals; wave transmission; multiple-scattering

1. INTRODUCTION

Photonic computing holds promise for superior processing of information in terms of speed, liability and bandwidth. Key ingredients in a photonic/optical computer are the waveguiding devices which steer light among different photonic components (diodes, optical transistor, optical buffers, *etc.*). Although this is easily accomplished in the microwave regime via conventional metallic waveguides, in the optical regime it still is a subject of intensive research as commercially available optical fibers cannot efficiently bend light around sharp corners at micrometer or even lower dimensions. Perhaps the most robust method for steering and bending light in the micrometer scale are the so-called photonic-crystal fibers [1]. Namely, light is localized efficiently within a dielectric or air core due to the suppression of propagation within the photonic-crystal cladding which exhibits an omnidirectional band gap for the waveguiding frequencies. Device performance in photonic-crystal fibers can be deteriorated by fabrication imperfections which lead to important backscattering of light, especially at sharp edges. A different approach for guiding light in photonic computers lies with the discipline of plasmonics wherein the excitation of surface plasmons in noble-metal components leads to light localization in the nanoscale enabling efficient light steering at very small volumes [2]. The main obstacle in this technology are the inherent losses of noble metals in the optical regime which degrade light guiding at sufficiently long distances.

A recent idea which has been put forward as an alternative to the current technologies for light guiding is the so-called unidirectional or one-way waveguiding, *i.e.*, propagation of light solely along one of the two possible directions in a straight waveguide. A direct consequence of the inhibition of light propagation in the other direction is the complete absence of light backscattering due to fabrication-induced disorder and imperfections or at very sharp guiding edges. This is easily achieved in the GHz regime via introducing time-reversal symmetry breaking in the microwave circuitry with the use of ferromagnetic materials and application of magnetic fields [3,4,5,6,7]. However, in the optical regime, again, it is really challenging to implement such a strategy as the magnetic contribution of ferromagnetic materials at this spectral region is very weak. Artificial structures made of magneto-optically responsive materials can potentially provide a means to achieve unidirectional waveguiding without however success up to now, due to their weak contribution to the overall electromagnetic response of such materials. Although the magneto-optical contribution can be very much enhanced with the insertion of plasmonic materials [8,9,10,11,12,13,14,15,16,17] , experimental evidence of unidirectional waveguiding via magneto-optical structures is still lacking.

On the other hand, there have been alternative approaches so as to achieve unidirectional propagation/guiding without the use of magnetic materials and subsequent application of an external magnetic field. Namely, unidirectional propagation has been reported in nonlinear photonic crystals (PCs) [18,19,20,21], in photonic-crystal heterostructures [22,23,24,25] as well as in 2D PCs with periodically modulated interfaces [26,27,28]. However, in the above cases, although the application of an external magnetic field is no longer needed, the range of frequencies and angles of incidence within which unidirectional propagation occurs is very narrow. Recently we have presented a linear, passive design of a purely dielectric 3D PC which possess unidirectional (one-way) photonic band gaps spanning over wide regions in both the Brillouin zone and EM spectrum [29]. The respective unit cell of the PC possess a degree of chirality while, at the same time, consists of different material scatterers giving rise to unidirectional wave propagation and frequency gaps. In the present work, we proceed further with this PC design and explore its potential for unidirectional light propagation. Namely, we consider PC designs with real materials simulating the response of colloidal photonic-crystal heterostructures. We show, in particular, that the emergence of unidirectional frequency bands and gaps is favored by materials with low refractive-index contrasts wherein the PC frequency bands are described as a perturbation to the free-photon frequency bands of an empty lattice.

2. DESCRIPTION OF THE PHOTONIC-CRYSTAL HETEROSTRUCTURE

We deal with the 3D photonic crystal shown in Figure 1. It can be described as a monoclinic crystal with a four-point basis. The crystal is viewed as a succession of planes of spheres parallel to the *xy*-plane. Each plane possesses the same 2D periodicity defined by the primitive vectors $\mathbf{a}_1 = (a, 0, 0)$ and $\mathbf{a}_2 = (0, a, 0)$ [(001) crystallographic surface]. This means that we have a square lattice with lattice vectors $\mathbf{R}_n = n_1\mathbf{a}_1 + n_2\mathbf{a}_2$, $n_1, n_2 = 0, \pm1, \pm2, \cdots$ and corresponding reciprocal lattice vectors $\mathbf{g} = m_1\mathbf{b}_1 + m_2\mathbf{b}_2$, $m_1, m_2 = 0, \pm1, \pm2, \cdots$ where $\mathbf{b}_i \cdot \mathbf{a}_j = 2\pi\delta_{ij}$, $i, j = 1, 2$. A unit layer of the crystal consists of four non-primitive planes of spheres at (0, 0, 0), (0, 0, $a/2$), (0, 0, a), and ($-0.3a$, 0, $3a/2$). All the spheres of the crystal of Figure 1 have the same radius $S = 0.2a$. The spheres of the first layer are of type A, of the second type B while the spheres at the third and fourth layers are of type C. The $(n + 1)$-th unit layer is obtained from the n-th layer by the primitive translation $\mathbf{a}_3 = (-0.3a, 0, 2a)$. Such a monoclinic PC lattice has only one mirror plane [xz-plane, *i.e.*, (010) crystallographic surface] and is therefore a lattice of low symmetry. We can thus expect that light transmission through a finite slab of such a monoclinic PC lattice would be different for incidence from the two opposite faces of the slab, if the slab is considered as a stack of planes parallel to a crystallographic direction other than the (010). In what follows, the different sphere types correspond to spheres made of different materials. In principle, one-way propagation can also be achieved by having spheres of different size or scatterers of dissimilar shape.

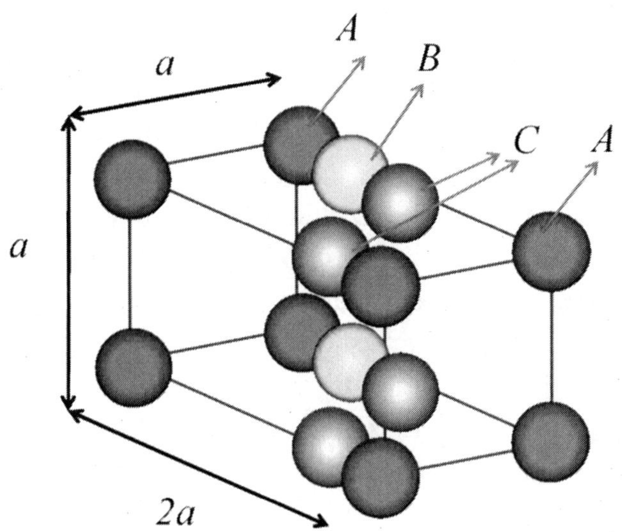

Figure 1. Unit cell of 3D PC with one-way photonic band gaps: monoclinic crystal consisting of four non-primitive planes of spheres parallel to the (001) surface at positions (0, 0, 0), (0, 0, a/2), (0, 0, a), and (−0.3a, 0, 3a/2).

3. CALCULATION METHOD

The PC considered in the preceding section will be studied in terms of its frequency band structure and light transmission/ absorption by employing the layer-multiple-scattering (LMS) method [30,31,32,33,34]. The LMS method is ideally suited for the calculation of the transmission, reflection and absorption coefficients of an EM wave incident on a composite slab consisting of a number of layers which can be either planes of non-overlapping spherical [30,31,32] or nonspherical axisymmetric [33] and of general shape [35] particles or clusters of such [34], with the same 2D periodicity or homogeneous plates. For each plane of particles, the method calculates the full multipole expansion of the total multiply scattered wave field and deduces the corresponding transmission and reflection matrices in the plane-wave basis. The transmission and reflection matrices of the composite slab are evaluated from those of the constituent layers. By imposing periodic boundary conditions one can also obtain the (complex) frequency band structure of an infinite periodic crystal. The method applies equally well to non-absorbing systems and to absorbing ones. Its main advantage over the other existing numerical methods lies in its efficient and reliable treatment of systems containing strongly dispersive materials such as Drude-like and polaritonic materials.

4. RESULTS AND DISCUSSION

In order to assess the asymmetrical transmission of light which would possibly lead to the generation of unidirectional photonic frequency bands/gaps, we calculate the transmittance of light incident on finite slabs of the 3D PC of Figure 1, based on the LMS method analyzed above. In Figure 2 we show a finite slab of the photonic crystals studied here wherein the slab consists of eight unit layers. As type-A spheres we have considered silica SiO_2 ($\epsilon = 2.1$) spheres, as type-B polystyrene ($\epsilon = 2.6$) spheres and as type-C silicon ($\epsilon = 11.9$) (Si) spheres. All three types of materials have been the basis for the fabrication of colloidal photonic crystals of nanospheres (for a review see [36]).

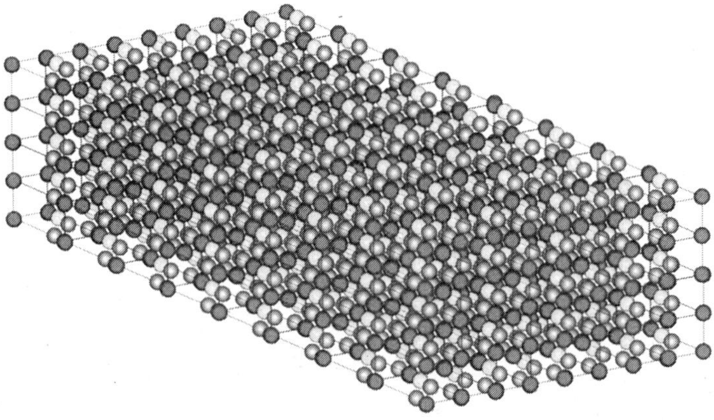

Figure 2. Finite slab of the photonic crystal of Figure 2 consisting of eight unit layers.

T^+ refers to the transmittance of light incident from the left (001) face of the PC slab, *i.e.*, a wave propagating from left to right whereas T^- refers to the transmittance of light incident from the right (001) face of the PC slab, *i.e.*, a wave propagating from right to left. Figure 3 shows T^+ and T^- for light incident off-normally on a finite slab of the Figure 2 with wavevector$\mathbf{k}_{\|} = (0.25, 0)\pi/a$, for s- (a) and p- (c) polarization. The PC slab consists of eight unit layers where each layer contains four square lattices (planes) arranged as shown in Figure 2. Evidently, there are spectral regions where T^+ and T^- differ significantly, such as, e.g., from $\omega a/c = 4.8 - 5.0$ and from $\omega a/c = 5.2 - 5.8$ for *s*-polarized light. For *p*-polarized light, T^+ and T^-differ substantially from $\omega a/c$= 5.9 – 6.1. Despite the important differences in light transmission for left and right incidence a clear case where there is a frequency gap for only one case of incidence direction (left or right) cannot be observed. By inspecting the corresponding frequency band structure (Figure 3b) we see that there exists a single band gap (from $\omega a/c = 5.0 - 5.2$) for both polarizations and directions of incidence (left and right) as well as other gaps for only *s*- or *p*-polarized incident light. We also observe a definite asymmetry in the frequency bands relative to the

$k_z = 0$ which stems from the lack of reflection symmetry at this crystal direction.

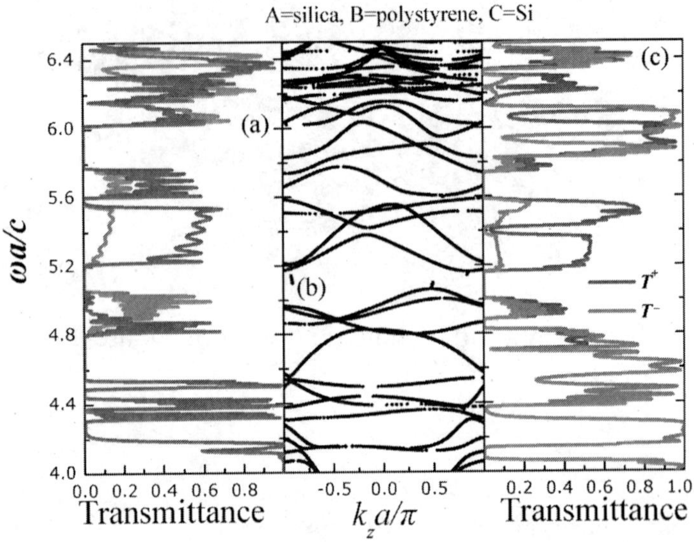

Figure 3. Transmittance for s- (a) and p- (c) polarized light incident with $\mathbf{k}_{\|} = (0.25, 0)\pi/a$ on a finite slab of the PC (see Figure 2) consisting of eight unit layers whereas type-A spheres we have considered silica SiO_2 ($\epsilon = 2.1$) spheres, as type-B polystyrene ($\epsilon = 2.6$) spheres and as type-C silicon ($\epsilon = 11.9$) (Si) spheres. T^+ (T^-) is the transmittance for light incident from the left (right) (001) faces of the slab. (b) Frequency band structure of the infinitely periodic PC of Figure 1 for $\mathbf{k}_{\|}= (0.25, 0)\pi/a$.

Next, we consider a PC heterostructure with the (same) unit cell of Figure 1 but with the polystyrene spheres (the B-type spheres in Figure 3) substituted with germanium ($\epsilon = 16.2$) ones. Evidently, both spectra of T^+ and T^- practically coincide apart from two very narrow regions denoted by the arrows in Figure 4c. This is in accordance with the corresponding frequency band structure of Figure 4b where all frequency bands are symmetric with respect to the center $k_z = 0$. Obviously, the inclusion of materials of high-index contrast in the considered photonic-crystal heterostructures does not favor the emergence of unidirectional propagation.

In an opposite trend (in relation with Figure 4), we substitute the high-index silicon spheres (type-C spheres) of Figure 4with sapphire ($\epsilon = 3.13$) ones [37]—see Figure 5. We clearly see an entirely different picture from that of Figure 4. The frequency bands show a clear asymmetry with respect to $k_z = 0$ which is reflected in the corresponding transmission spectra of Figure 5a,c. Namely, we identify two regions of clear unidirectional frequency gaps. Namely, for both *s*- and *p*-polarized light, there exists a T^+-frequency gap around $\omega a/c = 5.0$ and a T^--frequency gap around $\omega a/c = 5.4$. Asymmetric transmission (for left- and right-incidence) is also evidenced $\omega a/c = 5.8$. We note that the emergence of the unidirectional frequency gaps is not generated by complete absence of frequency bands for one of the k_z-directions ($k_z > 0$ or $k_z < 0$). There is a more complex mechanism which explains the emergence of unidirectional frequency gaps in terms of the matching of the group velocities of incident waves with that of the frequency bands (described analytically in [29]).

The comparison between Figure 4 and Figure 5 leads to the conclusion that the inclusion of high-index spheres does not favor the emergence of unidirectional gaps and propagation. This is a direct consequence of the localized nature of the EM modes of a PC consisting of high-index scatterers. Namely, due to the high refractive index (dielectric constant) of the scatterers the EM field is much localized within a plane of scatterers and wave propagation is achieved by hopping between neighboring planes in a manner resembling the tight-binding picture of electrons in ordinary atomic solids. In accordance with this picture, the lateral (normal to the growth direction of the crystal) displacement of a plane of spheres in an otherwise rectangular arrangement of planes of spheres, see, e.g., the fourth plane of C-type spheres of Figure 1, does not seem to affect seriously the frequency band structure

(Figure 4b) and subsequently the transmission spectrum. On the other hand, if the PC comprises of low-index spheres, the photons propagate nearly freely within the PC (a completely opposite picture than the tight-binding mechanism describing wave propagation in a PC of high-index scatterers) except from undergoing diffraction for the *k*-wavevectors implied by the underlying lattice symmetry. Since a monoclinic crystal structure yields an entirely different diffraction pattern than a rectangular one, we expect that the lateral displacement of the fourth plane (of C-type spheres) of Figure 1 to have a large impact on wave propagation of nearly free photons.

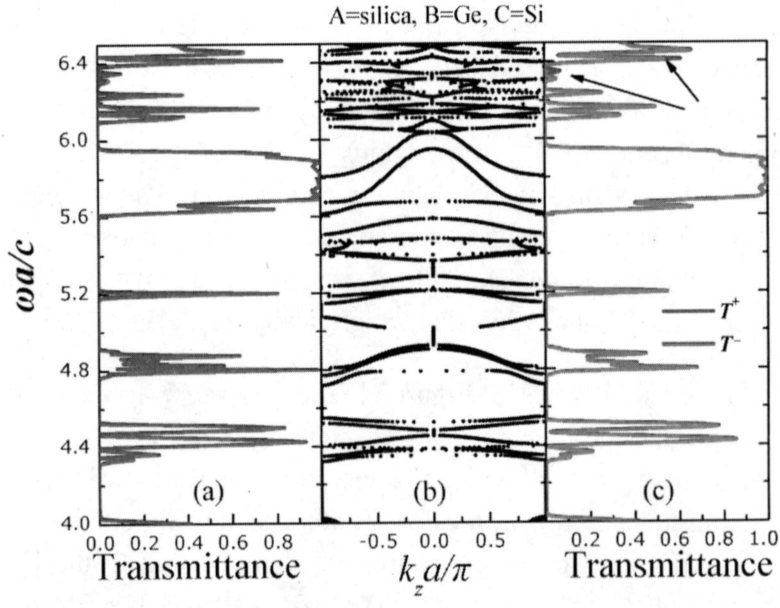

Figure 4. Transmittance for s- (**a**) and p- (**c**) polarized light incident with $\mathbf{k}_{\parallel} = (0.25, 0)\pi/a$ on a finite slab of the photonic crystal (PC) (see Figure 2) consisting of eight unit layers whereas type-A spheres we have considered silica SiO_2 ($\epsilon = 2.1$) spheres, as type-B germanium ($\epsilon = 16.2$) spheres and as type-C silicon ($\epsilon = 11.9$) (Si) spheres. T^+ (T^-) is the transmittance for light incident from the left (right) (001) faces of the slab; (**b**) Frequency band structure of the infinitely periodic PC of Figure 1 for $\mathbf{k}_{\parallel} = (0.25, 0)\pi/a$.

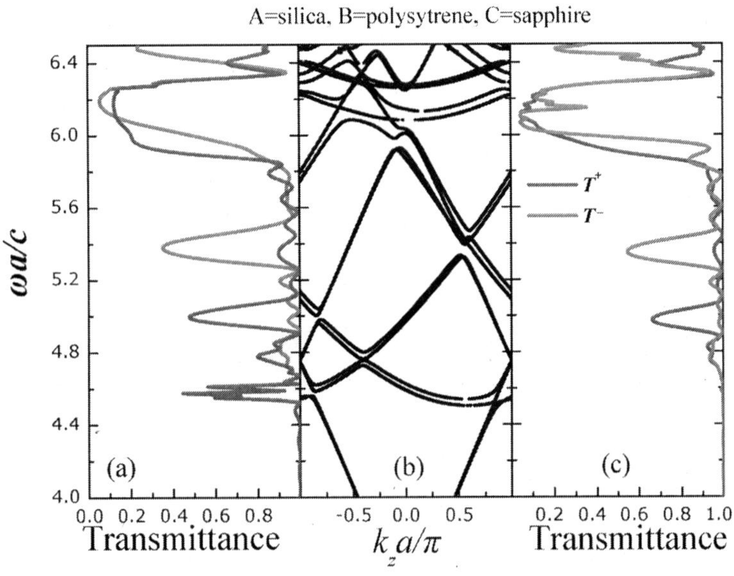

Figure 5. Transmittance for s- (**a**) and p- (**c**) polarized light incident with $\mathbf{k}_{||} = (0.25, 0)\pi/a$ on a finite slab of the PC of Figure 1 consisting of eight unit layers whereas type-A spheres we have considered silica SiO_2 ($\epsilon = 2.1$) spheres, as type-B polystyrene ($\epsilon = 2.6$) spheres and as type-C sapphire ($\epsilon = 3.13$) (Si) spheres. T^+ (T^-) is the transmittance for light incident from the left (right) (001) faces of the slab; (**b**) Frequency band structure of the infinitely periodic PC of Figure 1 for $\mathbf{k}_{||} = (0.25, 0)\pi/a$.

5. CONCLUSIONS

We have shown numerically that it is possible to obtain unidirectional EM wave propagation with plain, linear, dielectric photonic crystals consisting of colloidal micro- or nanospheres without applying external magnetic fields or using nonlinear materials. Prerequisites for observing unidirectional wave propagation are the low symmetry of the photonic crystals and the use of a heterostructure of photonic crystals of different scatterers (being different in material composition and/or size) as the unit cell. We have explored photonic crystals of different material composition wherein we have found that efficient unidirectional propagation in the form of unidirectional frequency gaps/bands emerges when scatterers of low-index contrast are used within the heterostructured unit cell of the crystal. For a practical realization, say, in the optical regime, of a low-symmetric photonic crystal such as those considered here, structures with lattice constant of about 480 nm are needed for a unidirectional band gap around 600 nm, which are usually fabricated via colloidal self-assembly [38,39,40,41].

CONFLICTS OF INTEREST

The authors declare no conflict of interest.

REFERENCES

1. Zolla, F.; Renversez, G.; Nicolet, A.; Kuhlmey, B.; Guenneau, S.; Felbacq, D. *Foundations of Photonic Crystal Fibres*; Imperial College Press: London, UK, 2005.

2. Bozhevolnyi, S. *Plasmonic Nanoguides and Circuits*; Pan Stanford Publishing: Singapore, 2009.

3. Haldane, F.D.M.; Raghu, S. Possible realization of directional optical waveguides in photonic crystals with broken time-reversal symmetry. *Phys. Rev. Lett.* **2008**, *100*, 013904.

4. Wang, Z.; Chong, Y.D.; Joannopoulos, J.D.; Soljacˇic´, M. Reflection-free one-way edge modes in a gyromagnetic photonic crystal. *Phys. Rev. Lett.* **2008**, *100*, 013905.

5. Wang, Z.; Chong, Y.D.; Joannopoulos, J.D.; Soljacˇic´, M. Observation of unidirectional backscattering-immune topological electromagnetic states. *Nature* **2009**, *461*, 772–775.

6. Yu, Z.; Veronis, G.; Wang, Z.; Fan, S. One-way electromagnetic waveguide formed at the interface between a plasmonic metal under a static magnetic field and a photonic crystal. *Phys. Rev. Lett.* **2008**, *100*, 023902.

7. Ao, X.; Lin, Z.; Chan, C.T. Dirac spectra and edge states in honeycomb plasmonic lattices. *Phys. Rev. B* **2009**, *80*, 033105.

8. Li, Y.; Zhang, Q.; Nurmikko, A.V.; Sun, S. Enhanced magnetooptical response in dumbbell-like Ag − $CoFe_2O_4$nanoparticle pairs. *Nano Lett.* **2005**, *5*, 1689–1692.

9. González-Díaz, J.B.; García-Martín, A.; García-Martín, J.M.; Cebollada, A.; Armelles, G.; Sepúlveda, B.; Alaverdyan, Y.; Käll, M. Plasmonic Au/Co/Au nanosandwiches with enhanced magneto-optical activity. *Small* **2008**, *4*, 202–205.

10. Jain, P.K.; Xiao, Y.; Walsworth, R.; Cohen, A.E. Surface plasmon resonance enhanced magneto-optics (SuPREMO): Faraday rotation enhancement in gold-coated iron oxide nanocrystals. *Nano Lett.* **2009**, *9*, 1644–1650.

11. González-Díaz, J.B.; Sepúlveda, B.; García-Martín, A.; Armelles, G. Cobalt dependence of the magneto-optical response in magnetoplasmonic nanodisks. *Appl. Phys. Lett.* **2010**, *97*, 043114.

12. Dani, R.K.; Wanh, H.; Bossmann, S.H.; Wysin, G.; Chikan, V. Faraday rotation enhancement of gold coated Fe2O3

nanoparticles: Comparison of experiment and theory. *J. Chem. Phys.* **2011**, *135*, 224502.

13. Pakdel, S.; Miri, M. Faraday rotation and circular dichroism spectra of gold and silver nanoparticle aggregates. *Phys. Rev. B* **2012**, *86*, 235445.

14. Armelles, G.; Cebollada, A.; García-Martín, A. ; Gonzá M.U. Magnetoplasmonics: Combining magnetic and plasmonic functionalities. *Adv. Opt. Mater.* **2013**, *1*, 10–35.

15. Christofi, A.; Stefanou, N. Nonreciprocal optical response of helical periodic structures of plasma spheres in a static magnetic field. *Phys. Rev. B.* **2013**, *87*, 115125.

16. Christofi, A.; Stefanou, N. Nonreciprocal photonic surface states in periodic structures of magnetized plasma nanospheres. *Phys. Rev. B.* **2013**, *88*, 125133.

17. Nahal, A.; Talebi, R. Ellipticity-dependent laser-induced optical gyrotropy in AgCl thin films doped by silver nanoparticles. *J. Nanopart. Res.* **2014**, *16*.

18. Soljacˇic´, M.; Luo, C.; Joannopoulos, J.D.; Fan, S. Nonlinear photonic crystal microdevices for optical integration. *Opt. Lett.* **2003**, *28*, 637–639.

19. Scalora, M.; Dowling, J.P.; Bowden, C.M.; Bloemer, M.J. Optical limiting and switching of ultrashort pulses in nonlinear photonic band gap materials. *Phys. Rev. Lett* **1994**, *73*, 1368–1371.

20. Mingaleev, S.F.; Kivshar, Y.S. Nonlinear transmission and light localization in photonic-crystal waveguides. *J. Opt. Soc. Am. B* **2002**, *19*, 2241–2249.

21. Yu, Z.; Fan, S. Complete optical isolation created by indirect interband photonic transitions. *Nat. Photon.* **2009,** *3*, 91–94.

22. Hwang, J.; Song, M.H.; Park, B.; Nishimura, S.; Toyooka, T.; Wu, J.W.; Takanishi, Y.; Ishikawa, K.; Takezoe, H. Electro-tunable optical diode based on photonic bandgap liquid-crystal heterojunctions. *Nat. Mater.* **2005**, *4*, 383–387.

23. Lu, C.; Hu, X.; Zhang, Y.; Li, Z.; Xu, X.; Yang, H.; Gong, Q. Ultralow power all-optical diode in photonic crystal heterostructures with broken spatial inversion symmetry. *Appl. Phys. Lett.* **2011**, *99*, 051107.

24. Wang, C.; Zhou, C.Z.; Li, Z.Y. On-chip optical diode based on silicon photonic crystal heterojunctions. *Opt. Express***2011**, *19*, 26948–26955.

25. Serebryannikov, A.E.; Nojima, S.; Ozbay, E. One-way absorption of terahertz waves in rod-type and multilayer structures containing polar dielectrics. *Phys. Rev. B* **2014**, *90*, 235126.

26. Lockyear, M.J.; Hibbins, A.P.; White, K.R.; Sambles, J.R. One-way diffraction grating. *Phys. Rev. E* **2006**, *74*, 056611.

27. Serebryannikov, A.E. One-way diffraction effects in photonic crystal gratings made of isotropic materials. *Phys. Rev. B***2009**, *80*, 155117.

28. Kuzmiak, V.; Maradudin, A.A. Asymmetric transmission of surface plasmon polaritons. *Phys. Rev. A* **2012**, *86*, 043805.

29. Yannopapas, V. One-way photonic band gaps and optical isolation with three-dimensional photonic crystals of low symmetry. *Phys. Rev. A* **2013**, *88*, 043837.

30. Stefanou, N.; Karathanos, V.; Modinos, A. Scattering of electromagnetic waves by periodic structures. *J. Phys. Condens. Matter* **1992**, *4*, 7389–7400.

31. Stefanou, N.; Yannopapas, V.; Modinos, A. Heterostructures of photonic crystals: Frequency bands and transmission coefficients. *Comput. Phys. Commun.* **1998**, *113*, 49–77.

32. Stefanou, N.; Yannopapas, V.; Modinos, A. MULTEM 2: A new version of the program for transmission and band-structure calculations of photonic crystals. *Comput. Phys. Commun.* **2000**, *132*, 189–196.

33. Gantzounis, G.; Stefanou, N.; Papanikolaou, N. Optical properties of periodic structures of metallic nanodisks. *Phys. Rev. B* **2008**, *77*, 035101.

34. Yannopapas, V.; Vanakaras, A.G. Layer-multiple-scattering theory for metamaterials made from clusters of nanoparticles. *Phys. Rev. B* **2011**, *84*, 085119.

35. Yannopapas, V. Layer-multiple-scattering method for photonic structures of general scatterers based on a discrete-dipole approximation/T-matrix point-matching method. *J. Opt. Soc. Am. B* **2014**, *31*, 631–636.

36. Kim, S.H.; Lee, S.Y.; Yang, S.M.; Yi, G.R. Self-assembled colloidal structures for photonics. *NPG Asia Mater.* **2011**, *3*, 25–33.

37. Ishii, M.; Kato, H.; Hashimoto, I.; Homma, Y. Synthesis of sapphire nanoparticles with graphite shells by hot-filament chemical vapor deposition. *Mater. Express* **2014**, *4*, 135–143.

38. Istrate, E.; Sargent, E.H. Photonic crystal heterostructures and interfaces. *Rev. Mod. Phys.* **2006**, *78*, 455–481.

39. Kedia, S.; Reddy, M.S.; Vijaya, R. Photonic crystal based direct and inverse heterostructures by colloidal self-assembly.*Opt. Photon. J.* **2012**, *2*, 242–248.

40. Cai, Z.; Liu, Y.J.; Teng, J.; Lu, X. Fabrication of large domain crack-free colloidal crystal heterostructures with superposition bandgaps using hydrophobic polystyrene spheres. *ACS Appl. Mater. Interface* **2012**, *4*, 5562–5569.

41. Liu, X.; Zhao, D.; Geng, C.; Zhang, L.; Tan, T.; Hu, M.; Yan, Q. Fabrication of colloidal photonic crystal heterostructures free of interface imperfection based on solvent vapor annealing. *J. Colloid Interface Sci.* 2014, *434*, 98–103.

CHAPTER 5

Analytical Study of the Propagation of Fast Longitudinal Modes along wz-BN/AlN Thin Acoustic Waveguides

Cinzia Caliendo

Istituto di Acustica e Sensori Corbino, IDASC-CNR,
Via del Fosso del Cavaliere 100, 00133 Roma, Italy

ABSTRACT

The propagation of the fundamental symmetric Lamb mode S_0 along wz-BN/AlN thin composite plates suitable for telecommunication and sensing applications is studied. The investigation of the acoustic field profile across the plate thickness revealed the presence of modes having longitudinal polarization, the Anisimkin Jr. plate modes (AMs), travelling at a phase velocity close to that of the wz-BN longitudinal bulk acoustic wave propagating in the same direction. The study of the S_0 mode phase velocity and coupling coefficient (K^2) dispersion curves, for different electrical boundary conditions, has shown that eight different coupling configurations are allowable that exhibit a K^2 as high as about 4% and very high phase velocity (up to about 16,700 m/s). The effect of the thickness and material type of the metal floating electrode on the K^2 dispersion curves has also been investigated, specifically addressing the design of an enhanced coupling device. The gravimetric sensitivity of the BN/AlN-based acoustic waveguides was then calculated for both the AMs and elliptically polarized S_0 modes; the AM-based sensor velocity and attenuation shifts due to the viscosity of a surrounding liquid was theoretically predicted. The performed investigation suggests that wz-BN/AlN is a very promising substrate material suitable for developing GHz band devices with enhanced electroacoustic coupling efficiency and suitable for application in telecommunications and sensing fields.

KEYWORDS

wz-BN; AlN; GHz range frequency; liquids; high coupling efficiency

1. INTRODUCTION

Electroacoustic devices based on the propagation of surface and bulk acoustic waves (SAW and BAW) are known for their wide range of applications in chemical, biological and physical sensing fields. When these devices are required to operate in a liquid environment, they must involve the propagation of specific types of acoustic waves that do not radiate energy into the liquid. Love waves, surface transverse waves (STW), shear horizontal acoustic plate modes (SHAPM) and shear horizontal SAW (SHSAW) are examples of acoustic waves whose shear horizontal polarization ensures no coupling between the liquid and the elastic propagating medium. Waves with elliptical polarization, such as the Rayleigh-like waves, are impractical for use in liquid environments as a consequence of the high energy loss in the liquid, both by radiating compressional waves into the fluid and due to frictional losses in the fluid. The fundamental antisymmetric Lamb mode, A_0, while being elliptically polarized, can travel along thin plates that are in contact with a liquid for a limited plate thickness range corresponding to mode velocities lower than the compressional velocity v_{liq} of the surrounding liquid medium. The fundamental symmetric Lamb mode, S_0, as well as the higher order symmetrical and antisymmetrical modes are elliptically polarized and have a velocity higher than that of the surrounding liquid medium, and thus, they are not suitable for sensing applications in liquids, except for some special cases. These cases include the linearly-polarized modes, such as the longitudinally-polarized Anisimkin Jr. modes (AMs), with a dominant longitudinal displacement component U_1 with a constant amplitude along the whole depth of the plate, the shear components, U_2 and U_3, being at least 10-times less than U_1 at any plate depth. AMs were first discovered in quartz plates [1] and then in $LiNbO_3$, Te [2], $LiTaO_3$, $Ba_2NaNb_5O_{15}$, KH_2PO_4, $Li_2B_4O_7$, TeO_2, $PbMoO_4$, KH_2PO_4 and ZnO [3,4] and, recently, in $GaPO_4$ [5]. The AMs, while having a phase velocity close to that of the longitudinal bulk acoustic wave (LBAW) traveling in the same direction, can be suitable for the development of high-frequency electroacoustic devices if implemented on fast materials, such as BN and AlN. Boron nitride (BN) is a group III–V compound semiconductor, and wurtzite (wz-BN) and cubic BN (zb-BN) are two representative crystal structures. The former is chemically inert and thermally stable and has been widely used as an electrical insulator and heat-resistant material. The latter, which is a super-hard material with properties similar to diamond, including chemical inertness, shows excellent properties, such as a high melting temperature, high thermal conductivity and chemical stability. zb- and wz-BN films are currently grown by sputtering (DC or RF) and by hollow cathode arc evaporation apparatus on Si(001) and Si(111), respectively [6–9]. AlN is the fastest piezoelectric material among those that can be grown in thin film form: compared to ZnO, for example, AlN films show higher SAW velocity (5607 m/s, as opposed to 2682 m/s), higher hardness (17.7 GPa, as opposed to 4.7 GPa) and thermal conductivity (2.8 $W{\cdot}cm^{-1}{\cdot}°C^{-1}$, as opposed to 0.6 $W{\cdot}cm^{-1}{\cdot}°C^{-1}$) and a slightly lower electroacoustic coupling coefficient K^2 (0.3%, as opposed to 0.97%). The AlN resistance to high temperature and to caustic chemicals make it the ideal candidate for the development of electroacoustic devices able to survive in harsh environment. The present paper investigates the acoustic wave propagation along wz-BN/AlN thin composite plates that are suitable for the development of high-frequency electroacoustic devices for telecommunications and sensing applications [10].

2. SAW AND BAW PROPAGATION ALONG BN AND ALN

The SAW phase velocity in zb-BN is highly affected by the material anisotropy, as well as the BAW; on the contrary, wz-BN and c-AlN are isotropic in the c-plane, and hence, both the SAW and BAW velocities are unaffected by the propagation direction in this plane. The SAW propagating along the AlN and wz-BN c-plane is a Rayleigh-type wave that has only two non-null rapidly-damped displacement components, U_1 and U_3, lying in the sagittal plane. The K^2 of the Rayleigh-like waves, as well as the velocities of the SAW and BAW propagating along the <100> direction of AlN and BN crystals are listed in Table 1. The acoustic wave velocity calculations were performed by using MATLAB software in the lossless approximation; the BN and AlN material data were deduced by [11,12].

Table 1. Electromechanical coupling and phase velocity of the Rayleigh-like SAW, phase velocities of the longitudinal (LBAW), shear vertical (SVBAW) and shear horizontal (SHBAW) bulk acoustic waves propagating along the <100> direction of AlN and wz-BN crystals.

Material	K^2 (%)	SAW (m/s)	LBAW (m/s)	SHBAW (m/s)	SVBAW (m/s)
c-AlN	0.28	5,607	10,287	5,814	6,089
wz-BN	0.04	9,719	16,781	11,027	10,587

As can be seen from Table 1, wz-BN is a weakly piezoelectric material showing very high SAW and BAW velocities, which make this material attractive for the development of high-frequency electro-acoustic devices.

3. ACOUSTIC WAVE PROPAGATION ALONG BN AND BN/ALN PLATES

3.1. BN Plates

Figure 1 shows the phase velocity of the fundamental symmetric Lamb mode, S_0, *vs.* the bare wz-BN plate thickness normalized to the wavelength, h_{BN}/λ. The dispersion curve is characterized by a low dispersion region, where the velocity is almost constant and very close to v_{LBAW}, followed by a drop in velocity. In the flat dispersion region, the S_0 mode transforms first to the AM counterpart with constant and dominant longitudinal displacement component U_1, while $U_2 = 0$ and $U_3 \leq 10\% \cdot U_1$. With increasing the plate thickness, the mode transforms to the quasi-longitudinal (QL) counterpart: the longitudinal displacement component is no longer constant through the thickness of the plate, but it is still dominant over the other two components ($U_2 = 0$ and $U_3 \ll U_1$). For higher plate thicknesses, when the S_0 mode velocity approaches the shear horizontal BAW velocity, the shear horizontal component U_2 largely dominates U_1 and U_3, but this wave is electrically uncoupled. The insets of Figure 1 show the three particle displacement components, U_1, U_2 and U_3, *vs.* the BN plate depth, for three different plate thicknesses, $h_{BN}/\lambda = 0.35$, 0.4 and 0.5. At $h_{BN}/\lambda = 0.35$, the QL mode (QLM) propagates, while with increasing h_{BN}/λ, the U_3 contribution slowly increases and becomes comparable to that of U_1, while the displacement component U_2 remains equal to zero. For $h_{BN}/\lambda > 1$, the mode begins to act as a Rayleigh wave propagating on the surfaces of the plate.

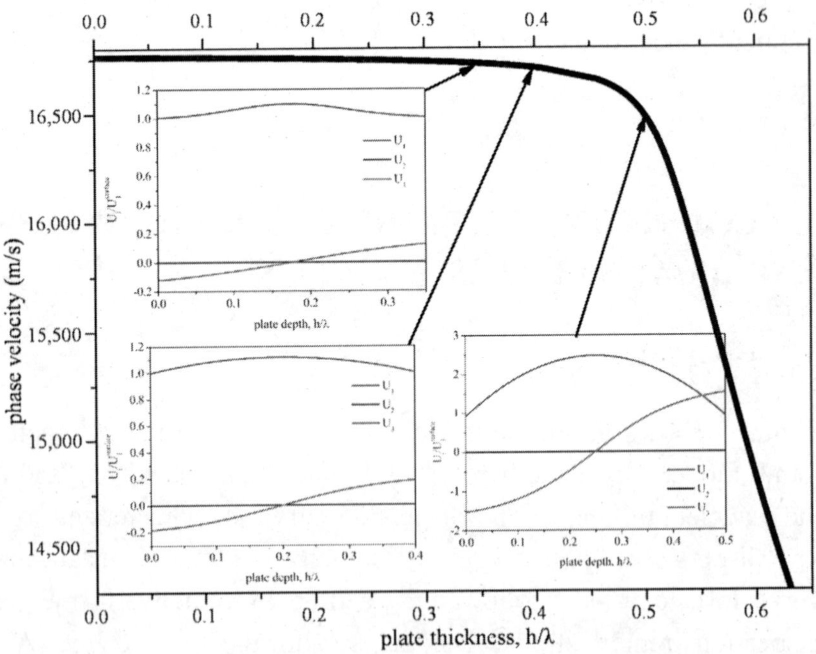

Figure 1. Phase velocity of the S_0 mode *vs.* h_{BN}/λ; the insets show the acoustic field profile for $h_{BN}/\lambda = 0.35$, 0.4 and 0.5.

The theoretical investigation of the mode field profile across the depth of the wz-BN thin bare plate revealed that the AMs propagate in the limited h_{BN}/λ range up to 0.325, with velocities from 16,760 ($h_{BN}/\lambda = 0.1$) to 16,735 m/s ($h_{BN}/\lambda = 0.325$).Figure 2 shows the depth profile of the AMs for four BN plate thicknesses: the displacement components U_1 and U_3 are normalized to the U_1 value at the surface. The shear horizontal component U_2 is not shown in Figure 2, as it is null. For $h_{BN}/\lambda > 0.325$, U_3 increases, and U_1 is no longer constant through the plate depth; the acoustic field profile of the S_0 mode changes according to the behavior shown in the insets of Figure 1.

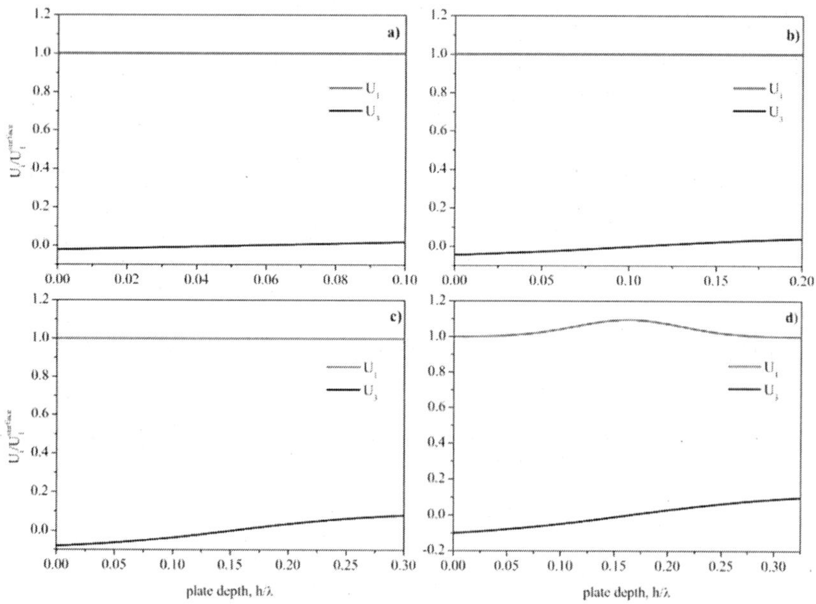

Figure 2. Depth profile of the AMs propagating along BN plates of different thicknesses normalized to the wavelength: **(a)** $h_{BN}/\lambda = 0.1$; **(b)** $h_{BN}/\lambda = 0.2$; **(c)** $h_{BN}/\lambda = 0.3$; and **(d)** $h_{BN}/\lambda = 0.325$.

Unfortunately, the piezoelectricity of BN is weak, thus BN-based electroacoustic devices are impractical without the addition of a thin piezoelectric film, such as AlN, whose technology is well established nowadays.

3.2. BN/AlN Composite Plates

The BN/AlN acoustic waveguide that we want to investigate consists of an acoustically thin suspended membrane that can be fabricated by a conventional bulk micromachining technique of a specifically orientated Si wafer. In addition to the Si substrate, the wz-BN thin film is to be used as an etch stop and to form the underlying membrane of the device. A piezoelectric AlN film is deposited on the BN layer, followed by a metallic layer (typically Al, Mo or Pt) on which the interdigital transducers (IDTs) are photolithographically patterned to generate and detect the propagating acoustic waves. The acoustic waves are generated by the application of an alternating voltage signal to the IDTs patterned on the piezoelectric substrate; the IDTs geometry dictates the wavelength of the excited wave, (λ = IDTs' period length), and the phase velocity of the wave dictates the device frequency, $f = v/\lambda$. The fabrication procedure of the BN/AlN-based acoustic waveguide is compatible with the semiconductor processing techniques, thus offering the advantage of providing the monolithic integration of the device with the signal processing electronics.

AMs propagate along BN/AlN composite plates only for very small AlN thicknesses that depend on the BN underlayer thickness. As an example, Figure 3 shows the AM field profile in the BN/AlN plate for $h_{BN}/\lambda = 0.1$ and $h_{AlN}/\lambda = 0.013$, 0.013 being the maximum AlN thickness that still allows the propagation of the AM through the composite plate with $h_{BN}/\lambda = 0.1$. As can be seen, the presence of the AlN layer weakly affects the AM depth profile, while for $h_{AlN}/\lambda > 0.013$, the AM converts into a QLM. For h_{AlN}/λ

< 0.013, the AM field profile along the plate depth is unaffected by the presence of the AlN layer, being $U_1 = 1$ across the plate depth. For $h_{BN}/\lambda = 0.2$, 0.3 and 0.325, the maximum h_{AlN}/λ values compatible with the AMs propagation along the BN/AlN composite plates are 0.01, 0.004 and 0.001, respectively.

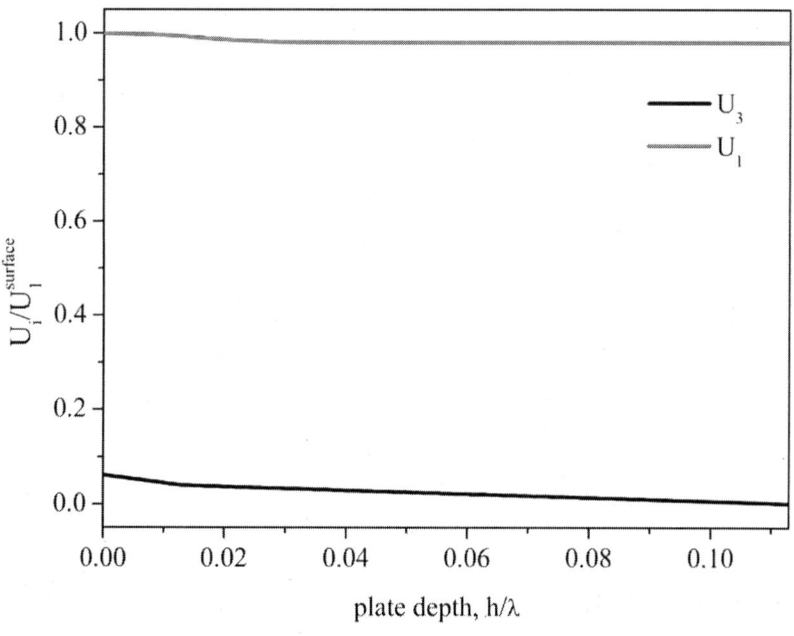

Figure 3. Depth profile of the AM propagating along the BN/AlN composite plate with $h_{BN}/\lambda = 0.1$ and $h_{AlN}/\lambda = 0.013$.

Figure 4 shows the S_0 mode dispersion curves, *i.e.*, the phase velocity *vs.* h_{AlN}/λ, along the BN/AlN composite plates, for different BN plate thicknesses ($h_{BN}/\lambda = 0.1$, 0.2, 0.3 and 0.325). As can be seen, for $h_{AlN}/\lambda \sim 0$, the velocity equals that of the bare BN plate, but increasing the h_{AlN}/λ, the mode phase velocity gradually decreases and asymptotically reaches the AlN SAW velocity for $h_{AlN}/\lambda > 1$. The presence of the AlN film lowers the mode velocity, but has the great advantage of increasing the coupling coefficient of the composite plate with respect to that of the bare BN plate, as shown in the next paragraph.

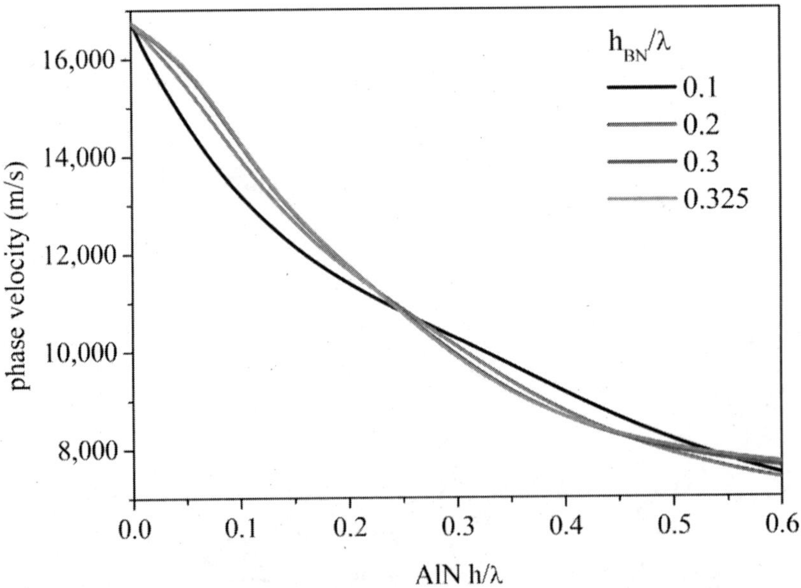

Figure 4. Phase velocity *vs.* AlN h/λ of the S_0 mode along the BN/AlN composite plates, h_{BN}/λ being the running parameter.

3.3. The Coupling Coefficient Dispersion Curves

The bare wz-BN plate is impractical for the development of an electroacoustic device, as its K^2 is weak, but this can be improved by covering the plate with a thin piezoelectric film, such as AlN, and selecting the proper electrical boundary conditions. The waveguide modes are excited and detected by the IDTs patterned on one surface of the AlN film: the application of an alternating electrical potential to one of the IDTs imposes an alternating strain field in the AlN film, which launches a mechanical wave into the composite plate. When the acoustic mode reaches the second transducer, it generates an electrical signal, which conveys information about the attenuation and phase delay experienced by the wave as it travels through the plate. By placing the IDTs on the AlN surface or at the BN/AlN interface, with or without one or two metallic layers opposite the IDTs, eight different electroacoustic coupling structures can be implemented, as depicted in Figure 5.

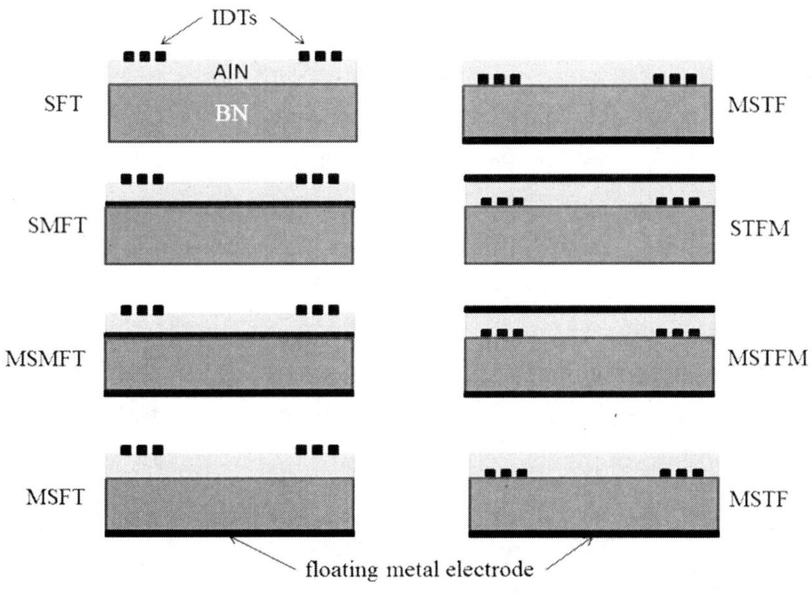

Figure 5. Cross-sections of the eight coupling configurations: the green, blue and black layers refer to the AlN (F, film), BN (S, substrate) and metal films (M, metal); the IDTs (T, transducers) are also shown.

The configuration called substrate/film/transducer (SFT) refers to a composite plate with the IDTs positioned on the AlN free surface: when a floating metallic plane (M, metal) is placed at the AlN/BN interface and/or at the BN free surface, the configurations are called SMFT, MSMFT or MSFT. The configuration called substrate/transducer/film (STF) refers to a composite plate with the IDTs positioned at the AlN/BN interface: when the metallic plane is positioned at the AlN free surface and/or at the BN free surface, the configurations are called STFM, MSTFM or MSTF. The K^2 is a measure of the electrical to acoustic energy conversion efficiency, and it was approximated to be $2[(v_f - v_m)/v_f]$, where v_m and v_f are the velocities calculated in the short-circuited and free boundary condition, in the approximation of an infinitesimally-thin perfectly-conductive floating electrode and IDTs. The K^2 dispersion curves were studied for the eight BN/AlN-based configurations, for different BN and AlN thicknesses normalized to the acoustic wavelength, h_{BN}/λ and h_{AlN}/λ, in the range 0.00001 to 0.6 with a step size of 0.0001. These curves were obtained taking into account only the electric loading from the IDTs and the metal floating plane. Figure 6 shows the K^2 *vs.* h_{AlN}/λ dispersion curves of the eight coupling structures, for $h_{BN}/\lambda = 0.1$, 0.2, 0.3 and 0.325: the insets show those K^2 dispersion curves that are not clearly distinguishable as a consequence of their partial overlapping with other curves. As can be seen, the K^2 of the BN/AlN plate depends very strongly on the electrical boundary conditions and on the BN and AlN layers thickness, and it reaches values higher than that of the single AlN piezoelectric film (0.28%).

For $h_{AlN}/\lambda \sim 0$, the K^2 of each configuration is small, but with increasing the AlN thickness, it increases and reaches a peak; for $h_{AlN}/\lambda \gg 1$, the K^2 moves toward the K^2 of the SAW that propagates in the (0001) plane of the AlN semi-infinite substrate. A K^2 as high as ~4% can be obtained at $h_{AlN}/\lambda \sim 0.27$ with the STFM and MSTFM structures on $h_{BN}/\lambda = 0.325$, for v~10,300 m/s. In general, the coupling configurations with the metal floating plane placed on one surface of the AlN layer (STFM, MSTFM) have a higher electromechanical coupling coefficient than that of the configurations with a free AlN surface (SFT, MSTF, MSFT), because the electric field strength through the AlN thin plate is higher than that between the IDT electrodes. The metallization on the AlN side opposite the IDTs strongly enhances the vertical electric field in the AlN layer.

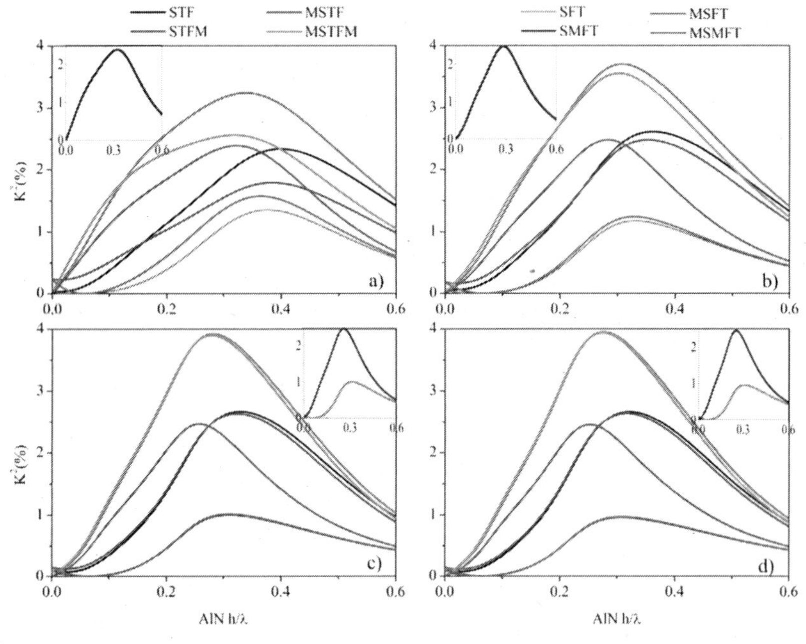

Figure 6. K^2 *vs.* h_{AlN}/λ dispersion curves of the eight BN/AlN-based coupling structures: **(a)**, **(b)**, **(c)** and **(d)** refer to $h_{BN}/\lambda = 0.1$, 0.2, 0.3 and 0.325, respectively.

The STFM and MSTFM coupling structures are the most efficient between the eight, and Table 2 lists the highest K^2 values that are obtainable by these two configurations for different BN/AlN thicknesses, together with the corresponding phase velocity.

All of the data listed in Table 2 refer to a Lamb-like wave with two non-null particle displacement components, U_1 and U_3; thus, such configurations are not suitable for operation in liquids, but only in a vacuum or air.

Table 2. The maximum K^2 value obtainable with the most efficient coupling structures, STFM and MSTFM, is listed together with the corresponding phase velocity and mass loading sensitivities, S and S′, evaluated with respect to the AlN and BN surfaces.

Configuration	$h_{BN}/\lambda / h_{AlN}/\lambda$	K^2_{max} (%)	v (m/s)	S (10^{-4} m²/Kgλ)	S' (10^{-4} m²/Kgλ)
STFM	0.1/0.34	3.24	9,754	−4.2	−0.87
	0.2/0.31	3.70	9,885	−5.0	−0.75
	0.3/0.28	3.92	10,204	−5.6	−0.66
	0.325/0.27	3.94	10,351	−5.8	−0.61
MSTFM	0.1/0.32	2.56	9,941	−4.5	−0.66
	0.2/0.3	3.55	10,024	−5.1	−0.61
	0.3/0.29	3.90	10,200	−5.6	−0.79
	0.325/0.274	3.94	10,284	−5.8	−0.67

As mentioned previously, the AMs can propagate along the BN/AlN composite plate only for a limited AlN thickness range that depends on the underlying BN layer thickness. Table 3 lists the highest AlN thicknesses that still allow the AMs' propagation on the BN/AlN plates, the K^2 and the velocity of the most efficient coupling configuration between the eight.

The AMs K^2 is lower than that of the Lamb modes of Table 2, but the velocity is greater and suitable for achieving very high operation frequencies (in the GHz band) with micron feature-sized IDTs.

All of the calculations listed in Tables 2 and 3 were performed taking into account only the electrical loading of the IDTs and the metal plane. A subsequent investigation accounting both for the electrical and mass loading of the IDTs and the metal electrode was performed; the effect of the metal electrode thickness and material type (Pt, Mo and W) was studied for the S_0 and AMs-based coupling structures listed in Tables 2 and 3. Different metal thickness ranges were studied relative to the AMs ($h_m/\lambda = 0.0001$ to 0.005, step size of 0.0001) and S_0 mode ($h_m/\lambda = 0.001$ to 0.01, step size of 0.001): the ranges were properly chosen, so as not to excite the propagation of unwanted acoustic modes. In the case of the AMs, the K^2 does not change with respect to the value calculated taking into account only the electric loading effect, but it decreases when h_m/λ is comparable with h_{AlN}/λ, whatever the metal type. In the case of the S_0 modes, a quite small improvement (2.65%) on K^2 can be obtained for the STFM configuration for the Mo electrode with thickness $h_m/\lambda = 0.007$ and for $h_{BN}/\lambda = 0.2$ to 0.325; the highest improvement in K^2 (from 7% to 10%) can be obtained for the MSTFM configurations with the W electrode, for $h_m/\lambda = 0.09$.

Table 3. The highest AlN thicknesses allowing the AM propagation, the K^2, the wave velocity and the gravimetric sensitivities, S and S′, of the most efficient coupling configurations, referring to the AlN and BN sensing side.

h_{BN}/λ / h_{AlN}/λ	K^2 (%)	v (m/s)	S (10^{-4} m²/Kgλ)	S' (10^{-4} m²/Kgλ)	Configuration
0.1/0.013	0.25	16,132	−12.30	−10.90	MSTFM
0.2/0.01	0.17	16,470	−6.92	−6.20	MSTF
0.3/0.004	0.14	16,664	−4.72	−4.40	MSTF
0.325/0.001	0.14	16,710	−4.20	−4.11	MSTF

4. BN/ALN-BASED ACOUSTIC WAVEGUIDES FOR SENSING APPLICATIONS

When an electroacoustic device is used for sensor applications, its output signal (a frequency and/or attenuation shift) is produced by the interaction between the acoustic wave (AW) and the chemical, biological or physical input to be measured. This interaction perturbs the AW characteristics, such as velocity and/or attenuation, through mechanical, acoustoelectric or viscoelastic coupling between the input stimulus to be measured and the acoustic wave field. For example, when an AW device is used for chemical vapor sensing, the wave propagation path is covered by a gas-specific sensing film: when this film is exposed to a gas, mechanical and electrical perturbations in the film will produce a shift in the resonant frequency of the acoustic wave device oscillator. Alternatively, by measuring the AW attenuation and the frequency shifts, one can determine the viscosity and density of a fluid that contacts these sensors. Generally the sensors based on electroacoustic devices are small, relatively inexpensive, quite sensitive and able to measure a wide variety of different input quantities [13].

4.1. Mass Sensors

The most common sensing application of the electroacoustic devices is based on the gravimetric principle for mass detection. A mass accumulation on the device surface changes the surface density of the propagating medium, hence resulting in a wave velocity shift. A mass sensor based on a high-frequency device exhibits superior performances: the higher the frequency, the higher the sensor sensitivity and resolution. If the added mass consists of an ideal thin elastic film that moves synchronously with the oscillating surface, the fractional velocity change-to-added mass ratio defines the sensor's gravimetric sensitivity S = [(v −

$v_l)/v_l]/m$, $m = \rho \cdot h$, ρ and h being the added layer's mass density and thickness, v and v_l the unloaded and mass-loaded plate's phase velocity. Tables 2 and 3 list the sensitivity data of the S_0 Lamb modes and the AMs propagating along the BN/AlN structures. Since the gravimetric detection can occur either on the AlN or BN plate surface, two normalized gravimetric sensitivity values, S and S′, were calculated for each AM and S_0-based configuration. S has been calculated as $S = [(v - v_0)/v_0]/(\rho h/\lambda)$, where v_0 and v are the wave velocities along the bare plate and the plate covered by a thin film, ρ is the film mass density and h/λ is the film thickness normalized to the wavelength. As can be seen, the AMs are more sensitive than the S_0 modes, and the AlN side of the composite plate is more sensitive than the BN side, irrespective of the thickness of both materials, the energy confinement at the BN side being lower than that at the AlN side. As an example, for λ = 10 μm, the AMs-based mass sensor (BN = 0.1λ = 1 μm, AlN = 0.013λ = 0.13 μm) shows S = −123 m^2/Kg, K^2 = 0.25% and f = 1613 MHz, and the S_0-based sensor (BN = 0.325λ = 3.25 μm, AlN = 0.274λ = 2.74 μm) shows S = −58 m^2/Kg, K^2 = 4% and f = 1028 MHz. These sensors show superior performances as compared to the SAW mass sensors implemented on STx-quartz, yx-, yz-and 128° tilted $LiNbO_3$, for which S = −36, −29, −21 and −31 m^2/Kg, K^2 = 0.12, 1.56, 4.82 and 4.60% and f = 316, 377, 340 and 380 MHz, respectively [14].

4.2. Viscosity Sensor

The AMs are a very attractive choice for sensing applications in liquids, as they have only one source of attenuation, the friction between the surface plate and the adjacent liquid, while the S_0 wave propagating in thick plates bordered by a liquid becomes a leaky wave due to the significant value of particle displacement normal to the plate surface and the radiation of compressional waves into the liquid. By monitoring the perturbations of the AMs propagation characteristics, such as the frequency and attenuation, one can measure the liquid density, viscosity, conductivity, permittivity and small mass changes. Although the AMs' gravimetric sensitivity was calculated in vacuum, it is a parameter of primary importance in the calculation of the sensitivity of a sensor for determining the viscosity of a fluid that contacts the device surface. The oscillation of the device surface contacting the liquid environment leads to entrainment of a thin liquid boundary layer near the interface that vibrates with a phase lag and an attenuated amplitude with respect to the surface vibration. The liquid layer velocity profile has an in-phase component that contributes to the viscous mass loading: a velocity shift arises from mass loading by the entrained viscous liquid layer of a thickness equal to the decay length of the AM field into the liquid, $\delta = [2\eta/\rho\omega]^{1/2}$, ρ and η being the liquid density and shear viscosity, $\omega = 2\pi f$ and $m' = [\rho_l\eta/2\omega]^{1/2}$ is the liquid mass that vibrates in phase with the plate mode. The velocity out-of-phase component contributes to the mode attenuation arising from the power dissipation in the liquid. The viscous coupling of the AMs to an adjacent Newtonian liquid results in both a change in the propagation velocity $\Delta v/v = (S/2\omega)\cdot[\eta\omega\rho_l/2]^{1/2}$ and attenuation $\alpha = (S/2v_0)\cdot[\omega\rho_l\eta/2]^{1/2}$ of the mode, and these changes are proportional to the mass sensitivity of plate mode velocity S [15,16]. As an example, for $\lambda = 100$ µm, at a frequency of 160 MHz and for 50% glycerin/water mixture ($\eta \cong 6.879$ mPa·s, $\rho \cong 1.144$ g/cm^3), $\delta \cong 0.1$ µm, $\Delta v/v_0 = 0.08\%$ and $\alpha = 0.34$ dB/cm, for the BN(0.1λ)/AlN(0.013λ)-based MSTFM configuration. This specific AM-based configuration, in addition to the advantage of having the highest mass sensitivity among the others, allows one to make measurements of the liquid viscosity that are not affected by the electric conductivity of the liquid that is being characterized, due to the presence of the metal electrode on the sensing side that short-circuits the electric field outside the plate. As the ability to detect perturbations on the plate side is affected by the materials' combination and the electrical boundary conditions, the measurements of the electrical conductivity of the liquid can be carried out using those configurations that provide for the direct contact between the liquid and the unmetallized surface of the BN/AlN-based configuration.

5. CONCLUSIONS

We report on the observation of very high velocity guided modes along thin BN/AlN composite plates, which exist when the plate thickness is about an order of magnitude smaller than the wavelength, while maintaining the electromechanical coupling coefficient of the piezoelectric AlN overlayer. The propagation characteristics of very fast longitudinal modes, the Anisimkin Jr. plate modes, along wz-BN/AlN thin composite plates is studied for different BN and AlN thicknesses and electrical boundary conditions. The study of the S_0 mode phase velocity and coupling coefficient K^2 dispersion curves revealed that eight different coupling configurations are allowable showing a K^2 as high as about 4% and a very high phase velocity (up to about 16,700 m/s). The gravimetric sensitivity of the BN/AlN-based configurations was theoretically predicted, specifically addressing the design of enhanced coupling, microwave frequency mass sensors based on AMs and S_0 modes, for applications in probing the solid/liquid interface and in a gaseous environment. The acoustic performances of these structures, such as the longitudinal polarization, the high K^2 and remarkable high phase velocity, suggests the novelty and the advantages of the AlN and BN combination, suitable for the development of a small, completely integrated sensor for monitoring the liquid properties. BN/AlN is a very promising substrate material suitable for developing GHz-band devices with enhanced electroacoustic coupling efficiency and suitable for working in a liquid environment.

CONFLICTS OF INTEREST

The author declares no conflict of interest.

REFERENCES

1. Anisimkin, I.V. New type of an acoustic plate mode: Quasi-longitudinal normal wave. *Ultrasonics* **2004**, *42*, 1095–1099.

2. Gulyaev, Y.V. Propagation of the Anisimkin Jr.' plate modes in $LiNbO_3$ and Te single crystals. Proceedings of the 2008 IEEE International Ultrasonics Symposium (IUS), Beijing, China, 2–5 November 2008; pp. 2221–2224.

3. Anisimkin, V. General Properties of the Anisimkin Jr. Plate Modes. *IEEE Trans. Ultrason. Ferroelectr. Freq. Control* **2010**,*57*, 2028–2034.

4. Anisimkin, V.I.; Pyataikin, I.I.; Voronova, N.V. Propagation of the Anisimkin Jr.' and Quasi-Longitudinal Acoustic Plate Modes in Low-Symmetry Crystals of Arbitrary Orientation. *IEEE Trans. Ultrason. Ferroelectr. Freq. Control* **2012**,*59*, 806–810.

5. Caliendo, C.; lo Castro, F. Quasi-linear polarized modes in y-rotated piezoelectric $GaPO_4$ plates. *Crystals* **2014**, *4*, 228–240.

6. Hu, C.; Kotake, S.; Suzuki, Y.; Senoo, M. Boron nitride thin films synthesized by reactive sputtering. *Vacuum* **2000**, *59*, 748–754.

7. Deng, J.-X.; Zhang, X.-K.; Yao, Q.; Wang, X.-Y.; Chen, G.-H.; He, D.-Y. Optical properties of hexagonal boron nitride

thin films deposited by radio frequency bias magnetron sputtering. *Chin. Phys. B* **2009**, *18*, 4013–4018.

8. Chen, X.; Sun, L.; Yang, B.; Guo, Y.; Wu, X. Research on the piezoelectric response of cubic and hexagonal boron nitride films. *Optoelectron. Lett.* **2012**, *8*, 117–120.

9. Sun, L.; Chen, X. Research on the h-BN films for high frequency SAW devices. Proceedings of the SPIE 8202 International Conference on Optical Instruments and Technology: Solid State Lighting and Display Technologies, Holography, Speckle Pattern Interferometry, and Micro/Nano Manufacturing and Metrology, Beijing, China, 28 November 2011; pp. 82020L:1–82020L:7.

10. Anisimkin, I.V.; Anisimkin, V.I. Attenuation of Acoustic Normal Modes in Piezoelectric Plates Loaded by Viscous Liquids. *IEEE Trans. Ultrason. Ferroelectr. Freq. Control* **2006**, *53*, 1487–1492.

11. Shimata, K.; Sota, T.; Suzuki, K. First principles study on electronic and elastic properties of BN, AlN and GaN. *J. Appl. Phys.* **1998**, *84*, 4951–4958.

12. *Landolt-Börnstein, Group III*; Springer-Verlag: Berlin, Germany, 1979; Volume 11.

13. Ballantine, D.S.; White, R.M.; Martin, S.J.; Ricco, A.J.; Zellers, E.T.; Fye, G.C.; Wohltjen, H. *Acoustic Wave Sensors: Theory, Design, & Physico-Chemical Applications*; Academic Press INC.: San Diego, CA, 1997.

14. Caliendo, C. Theoretical investigation of high velocity, temperature compensated Rayleigh waves along AlN/SiC substrates for high sensitivity mass sensors. *Appl. Phys. Lett.* **2012**, *100*. doi.org/10.1063/1.3675619.

15. Wu, J.; Zhu, Z. Sensitivity of Lamb Wave Sensors in Liquid Sensing. *IEEE Trans. Ultrason. Ferroelectr. Freq. Control* **1996**, *43*, 71–72.

16. Nayfeh, A.H.; Nagy, P.B. Excess attenuation of leaky Lamb waves due to viscous fluid loading. *J. Acoust. Soc. Am.* 1997, *101*, 2649–2658.

CHAPTER 6

A Scalar Acoustic Equation for Gases, Liquids, and Solids, Including Viscoelastic Media

Eugen Mamontov[1*], Viktor Berbyuk[2]

[1] Utilization Group, Department of Research and Development, Foundation Chalmers Industrial Technology, Gothenburg, Sweden

[2] Division of Dynamics, Department of Applied Mechanics, Chalmers University of Technology, Gothenburg, Sweden

ABSTRACT

The work deals with a mathematical model for real-time acoustic monitoring of material parameters of media in multi-state viscoelastic engineering systems continuously operating in irregular external environments (e.g., wind turbines in cold climate areas, aircrafts, etc.). This monitoring is a high-reliability time-critical task. The work consistently derives a scalar wave PDE of the Stokes type for the non-equilibrium part (NEP) of the average normal stress in a medium. The explicit expression for the NEP of the corresponding pressure and the solution-adequateness condition are also obtained. The derived Stokes-type wave equation includes the stress relaxation time and is applicable to gases, liquids, and solids.

KEYWORDS

Acoustic Monitoring, Gas, Liquid, or Solid, Acoustic Equation, Visoelastic Media, Stress Relaxation Time, Average Normal Stress, the Stokes-Type Wave Equation

1. INTRODUCTION

One of the applications of the acoustic-sensing technology is monitoring of material parameters of engineering systems continuously operating in irregular external environments. This type of the operation indicates that the monitoring must be regular (e.g., periodic) and in the real-time mode. Many problems in this area deal with the systems that are multi-state and viscoelastic in the following sense. A system comprises at least two spatial domains, each of which is occupied with an isotropic medium that is spatially homogeneous at equilibrium and is at one of the three states of matter: gaseous, liquid, or solid. In addition to that, the states of at least two of these spatial components are different, and the components are generally visoelastic.

The features of the considered systems and available mathematical models used for the system material parameter monitoring are further specified and discussed in more detail below.

a) The regular monitoring of material parameters is implemented by the non-invasive sensing of acoustic signals in one or more components of the system. The subsequent identification of the parameters is performed by using the sensed signals and the corresponding medium-specific acoustic models (e.g., [1] [2]).

b) The regular real-time acoustic monitoring presents the sequence of the sensing cycles started at a series of time points and implemented with one or more sensors in the automatic mode. If, say, $t'$and $t'' > t'$ are any two consecutive time points, the model-based processing of the sensed data received from all of the sensors must be completed unconditionally (e.g., without human intervention) during the time interval of the length $t'' - t'$. Otherwise, the data sensed in the t'-cycle will not be processed and, thus, will not contribute to the material parameter identification.

Moreover, the aforementioned continuous automatic operation presumes zero user intervention. This makes high demands of reliability of the data processing. The above picture indicates that the regular real-time moni- toring is a high-reliability time-critical task.

c) Due to the above multi-state feature of the system, the models mentioned in Point (a) generally include acoustic models for fluids and solids. In each of these cases, they are not formulated for acoustic signals, i.e. non-equilibrium parts of the Cauchy stress matrix entries. The fluid acoustic models are formulated for the entries of the velocity vector, whereas the solid acoustic models are formulated for the entries of the displace- ment vector. This inevitably complicates the entire description necessary for the parameter identification. More- over, the diversity and complexity of the modeling are further contributed by the use of representations for the conjunction of the initial conditions and the boundary conditions at the interfaces between the system com- ponents, which are at different states of matter (see above).

d) Acoustic models for fluids and solids generally include a system of three scalar non-stationary partial differential equations (PDEs) in the three-dimensional physical space.

e) Common fluid mechanics acoustic models natively include not only elastic moduli of the medium but also its viscosities, and thereby they are applicable to the corresponding viscoelastic components of the system. Also, the mentioned models are consistently derived from a more general physical theory, kinetic theory (e.g., [3]) which, in turn, results form statistical mechanics (e.g., [4]).

In contrast to that, common solid mechanics acoustic models include elastic moduli but do not include vis- cosities. There is an advanced model for visoelastic solids (e.g., [5] , (6.15)). It is based on the stress expression ([5] , (6.14)), which is theoretically explained in ([6] , §34), and includes both volume and shear viscosities of the medium. It appears to be a system of the Stokes-type wave PDEs. The term "Stokes-type" is due to the work of G.G. Stokes [7].

However, as follows from the discussion in ([6] , §34] (see also [5], the text on (6.9), (6.14), and (6.15)), the advanced PDE system is a compound model. It is obtained by means of adding of the viscous stress to the inviscid/elastic stress. Thus, it is not derived consistently from more general physical theories. In this respect, the basis of the model has a significant heuristic content. As opposed to that, the Stokes equation [7] was derived consistently within theory of viscous fluids for the velocity potential.

The features of common acoustic continuum mechanics models listed in Points (c)-(e) are not well suited for the time-critical nature indicated in Point (b). Consequently, a modeling basis for regular real-time acoustic monitoring of material parameters of multi-state engineering systems continuously operating in irregular exter- nal environments remains a research topic. The purpose of the present work is to contribute to this topic. More specifically, the work derives an acoustic PDE for an appropriate scalar component of the Cauchy stress matrix and explains why this PDE is applicable to gases, liquids, and solids, including viscoelastic media.

It should be noted that the idea of PDEs for the entries of the Cauchy stress matrix goes back to at least H. Grad who derives non-stationary spatially non-homogeneous PDE system for these entries ([3] , (28.19)). (The version of the system in the spatially homogeneous case is [3], (28.22).) Nowadays, PDEs for the entries of the Cauchy stress matrix (or its other components) in fact form a new area in acoustic modeling because they allow concentrating attention directly on the quantities of the main interest in acoustics, without involvement of inter- mediate variables (such as the displacement vector in the case of solids or the velocity vector in the case of vis- cous fluids). This direction was contributed by other works. For instance, work [8] formulates the acoustic PDEs for all of the aforementioned entries in solids ([8] , the equation in the article “0138”). This feature allows [8] better focusing the models on sharp practical applications.

The work is arranged as follows. Section 2 summarizes the basic facts on the Cauchy stress matrix and the key component of it, scalar and deviatoric stresses. The main result derived in the work is presented in Section 3 that also discusses the novelty of it and its connection to the related results of other authors. Section 4 concludes the work. The detailed derivation of the main result is carried out in Appendix A. It applies selected represent- ations associated with the coupling of Eulerian and Lagrangian coordinates outlined in Appendix B.

2. PRELIMINARIES: SCALAR AND DEVIATORIC STRESSES

Acoustic signals present the spatiotemporal deviation, ΔS, of the non-equilibrium part (NEP) of the Cauchy stress matrix, S, of the medium from its equilibrium version $\overline{S}$,

$$\Delta S = S - \overline{S}. \tag{2.1}$$

The terms denoted with the sign "overline" are specified in the remark below.

Remark 2.1. As is well known, physical quantities at equilibrium are independent of time. The present work considers the media only such that, at equilibrium, they are independent of space as well. Consequently, the equilibrium versions of physical quantities do not depend on space either. These versions are denoted with the sign "overline" applied to the notation of the corresponding quantity (e.g., see (2.1)). ,

One usually represents matrix S in the form of two components,

$$S = PI + Z \tag{2.2}$$

where

$$P = (1/3)\,\mathrm{tr}(S), \tag{2.3}$$

I is the identity 3×3-matrix, and $\mathrm{tr}(\cdot)$ is the trace of a matrix. As (2.3) shows, P is a scalar variable and, thus, matrix PI in (2.2) presents the scalar stress.

The diagonal and off-diagonal entries of matrix S are known as the scalar normal and shear stresses, re- spectively. Since (2.3) determines P as the arithmetic mean of the total normal stresses, P is called the average normal stress (ANS).

As follows from (2.2) and (2.3), matrix Z is traceless, i.e. such that $\mathrm{tr}(Z)=0$. For this reason, matrix

$$Z = S - PI \tag{2.4}$$

is called the deviatoric stress. Also, this stress is zero at equilibrium, i.e.

$$\bar{Z} = 0. \tag{2.5}$$

The relaxation of deviatoric stress Z to its equilibrium value (2.5) is usually described in terms of the stress relaxation time, say, θ, and according to asymptotic representation $\mathrm{d}Z/\mathrm{d}t = -Z/\theta$. The stress relaxation exists in any material medium, in gases, liquids, and solids, no matter if the medium is spatially non-homogeneous or spatially homogeneous (e.g., see both [3] , (28.19) and [3] , (28.22) for the case of fluids).

In an isotropic medium, deviatoric stress Z explicitly depends on shear modulus G of the medium, no matter if the latter is a solid (e.g., [5] , (1.43)) or a fluid (e.g., [9] , [10] , p. 655). In the case of a fluid, this depen- dence is presented implicitly, by means of the explicit dependence on shear viscosity of the medium $\mu = G\theta$.

For the sake of simplicity, we also use θ in the expression $\eta = K\theta$ for volume viscosity η where K is the bulk modulus of the medium. Deviatoric stress Z does not depend on bulk modulus K. The equilibrium versions of the mentioned equality is

$$\bar{\eta} = \bar{K}\bar{\theta}. \tag{2.6}$$

Value $\bar{K}$ and the equilibrium value $\bar{\rho}$ of the mass density of the medium ρ determine parameter

$$\bar{s} = \sqrt{\bar{K}/\bar{\rho}}. \tag{2.7}$$

This parameter is sometimes called the speed of bulk waves.

We note that ANS P completely determines not only scalar stress PI but also the entire stress S at equilibrium with expression

$$\bar{S} = \bar{P}I \tag{2.8}$$

that follows from (2.2) and (2.5). In view of of (2.2), (2.8), and (2.5), Expression (2.1) is equaivalent to

$$\Delta S = \Delta PI + Z \tag{2.9}$$

where

$$\Delta P = P - \bar{P}. \tag{2.10}$$

Remark 2.2. If the medium is close to the equilibrium state sufficiently in order to neglect deviatoric stress Z in (2.2), then (2.2) and (2.9) are reduced to $S = PI$ and

$$\Delta S = \Delta PI \tag{2.11}$$

respectively,

Some of the above relations are used in the derivation of the main result of the present work (see Appendix A).

3. THE STOKES-TYPE WAVE PDE FOR THE NON-EQUILIBRIUM PART OF THE AVERAGE NORMAL STRESS

As is shown in Appendix A, under the assumptions listed in Table 1, a closed description for the NEP ΔP of ANS P is PDE

$$\partial^2 \Delta P/\partial t^2 = \overline{s}^2 \nabla^2 \left(\Delta P + \overline{\theta}\, \partial \Delta P/\partial t\right). \tag{3.1}$$

The derivation of (3.1) also provides the corresponding description for the NEP Δp of pressure p (see (A.1.11)), which is any of the following two relations

$$\Delta P = \Delta p + \overline{\theta}\, \partial \Delta p/\partial t, \tag{3.2}$$

$$\Delta p = \Delta P + \overline{\theta}^2 \overline{s}^2 \nabla^2 \Delta P - \overline{\theta}\, \partial \Delta P/\partial t. \tag{3.3}$$

Equation (3.1) and any of (3.2) and (3.3) are linear. The corresponding solution-adequateness condition is (A.3.7) (see Proposition A.3.1).

In comparison with common wave PDE $\partial^2 \Delta P/\partial t^2 = \overline{s}^2 \nabla^2 \Delta P$, PDE (3.1) includes an extra term, the one with stress relaxation time $\overline{\theta}$. This is a damping term. It, however, does not result from the choice of one or another damping model. On the contrary, it is consistently derived from the basic laws for continuum media (see Appendix A). The damping term represents the internal, viscous friction in the medium.

Remark 3.1. The derivation of model (3.1)-(3.3) follows theory of viscous fluids but admits the terms native in theory of inviscid solids (see Remarks A.1.1 and A.2.1). Thus, the derived model is suitable for gases, liquids, and solids.

Table 1. Assumptions used in the derivation of acoustic equation system (3.2), (A.3.6).

1	The medium is spatially homogeneous and isotropic at equilibrium.
2	Elastic properties of the medium can be treated in terms of linear elasticity.
3	The medium is close to the equilibrium state sufficiently in order to neglect deviatoric component (2.4) of the total stress (2.2) (see also (2.5)).
4	The medium is assumed to be isothermal. As is well known (e.g., [16], p. 617), acoustical vibrations are almost always so rapid that there is no time for conduction to remove the heat developed and equalize the temperatures. The contractions and expansions take place adiabatically, *i.e.* without loss of heat. In spite of that, the above assumption on the isothermalness is used. The reason is avoiding the need in description of the spatiotemporal evolution of the temperature in the medium and, thereby, keeping the complexity of the model at a reasonable level. This in particular means that the aforementioned heat is neglected.
5	There are no chemical reactions in the medium.
6	There are no body forces in the medium.
7	If the medium is not a linear solid, then inequality (A.1.13) holds. According to Proposition A.3.1, this inequality can be replaced with (A.3.7).
8	The medium is close to the equilibrium state sufficiently in order to replace η in (A.3.2) with its equilibrium value $\bar{\eta}$.
9	The medium is close to the equilibrium state sufficiently in order to neglect velocity v in expression (A.3) thereby reducing it to (B.5).

We also note a connection of PDE (3.1) to a special wave equation that was introduced in 1845. Formally, PDE (3.1) for NEP ΔP of ANS P is identical to the Stokes wave PDE [7] for the velocity potential in a viscous fluid. Thus, (3.1) is one of the Stokes-type wave PDEs.

The Stokes-type wave PDEs for different variables are used in acoustic of viscoelastic solids since long ago. For example, Section 1 discusses the well-known Stokes-type wave PDE system for the displacement vector in a solid. Paper [11] analyzes propagation of plane and spherical waves in viscoelastic solids with the help of the normalized Stokes-type wave PDE, which is mathematically equivalent to (3.1). Book [12] discusses transient waves in gases, liquids, and solids in connection with applications to viscoelastic-solid acoustics in seismology. The entire modeling in this book is based on the Stokes wave PDE but treats its variable in a broader sense, as a generating function (e.g., [12] , (31) on p. 36). The in-depth discussion in ([12] , the chapter "Epilogue") em- phasizes a number of the advantages of the Stokes-type models.

4. CONCLUDING REMARKS

The present work considers material media, which are isothermal, spatially homogeneous and isotropic at equili- brium, with elastic properties treatable in terms of linear elasticity, and can be gaseous, liquid, or solid. The che- mical reactions and body forces in the media are neglected.

Under the assumptions listed in Table 1, the work consistently derives a scalar wave PDE (3.1) for the NEP of the ANS in the medium. Normal stress in any medium turns up in almost all situations, dynamic or not. Equation (3.1) appears to be a wave PDE of the Stokes type. It is endowed with the explicit expression for the NEP of the pressure, namely any of (3.2) and (3.3), and the solution-adequateness condition (A.3.7). This condition enables to check the adequateness of solutions of PDE (3.1). The derived model is applicable to media at different states of matter: gaseous, liquid, or solid (see Remark 3.1).

Application of the derived equation allows to overcome the difficulties emphasized in Points (c)-(e) in Section 1 and thereby meet requirements resulting from the high-reliability and time-critical nature formulated in Point (b) in the mentioned section.

ACKNOWLEDGEMENTS

Eugen Mamontov, Viktor Berbyuk The authors express their gratitude to the Swedish Energy Agency Project 37286-1 for a partial support of the present work. The authors also thank Anders Boström, the Head of the Division of Dynamics, Department of Applied Mechanics, Chalmers University of Technology, Gothenburg, Sweden, for a stimulating discussion.

REFERENCES

1. Rose, J.L. (1999) Ultrasonic Waves in Solid Media. Cambridge University Press, Cambridge.

2. Datta, S.K. and Shah, A.H. (2009) Elastic Waves in Composite Media and Structures: With Applications to Ultrasonic Nondestructive Evaluation. CRC Press, Boca Raton.

3. Grad, H. (1958) Principles of the Kinetic theory of Gases. In: Flügge, S. Ed., Handbuch der Physik, Band XII, Springer-Verlag, Berlin, 205-294.

4. Balescu, R. (1997) Statistical Dynamics: Matter out of Equilibrium. Imperial College Press, London.

5. Pollard, H.F. (1977) Sound Waves in Solids. Pion, London.

6. Landau, L.D. and Lifshitz, E.M. (1986) Theory of Elasticity. Pergamon Press, Oxford.

7. Stokes, G.G. (1845) On the Theories of the Internal Friction of Fluids in Motion and of the Equilibrium and Motion of Elastic Solids. Trans. Cambridge Philos. Soc., 8, 287-319.

8. Merkel, H. (2006) Apparatus and a Method for Determining the Spatial Distribution of Physical Parameters in an Object. US Patent Application 11/375,133, 20.

9. Landau, L.D. and Lifshitz, E.M. (1987) Fluid Mechanics. Pergamon Press, Oxford.

10. Koshlyakov, N.S., Smirnov, M.M. and Gliner, E.B. (1964) Differential Equations of Mathematical Physics. North-Holland Publishing, Amsterdam.

11. Longman, I.M. (1980) Wave Propagation in a Viscoelastic Solid. Journal of Computational Physics, 37, 171-182.

12. Ricker, N.H. (1977) Transient Waves in Visco-Elastic Media. Elsevier, Amsterdam.

13. Sedov, L.I. (1971) A Course in Continuum Mechanics. Vol. 1, Wolters-Noordhoff, Groningen.

14. Trigg, G.L. (1991) Encyclopedia of Applied Physics, Vol. 1, VCB, New York.

15. Dukhin, A., et al. (2014) Volume Viscosity, Wikipedia. Wikimedia Foundation, Inc.

16. Goldstein, H. (1980) Classical Mechanics. Addison-Wesley, Reading.

CHAPTER 7

Influence of the Chemical Composition of Completion Fluids on the Propagation of Electromagnetic Waves within Oil Wells

Alexandre Ashade Lassance Cunha, Marco Aurélio Pacheco, José Ricardo Bergmann

Departamento de Engenharia Elétrica, Pontifícia Universidade Católica do Rio de Janeiro, Rio de Janeiro, Brazil

ABSTRACT

The propagation of electromagnetic waves in the annular region of oil wells was studied. The present study aims to analyse the propagation attenuation along the well, as well as the input impedance determined by a source placed near the wellhead. A coaxial waveguide model was adopted with heterogeneous dielectrics and losses. First, a wave equation solution for the waveguide is presented, assuming a homogeneous medium with losses, by solving the equation in cylindrical coordinates using the vector potential technique. An uncertainty analysis model is then developed to model the heterogeneous characteristics of the medium. Monte Carlo simulations were performed with the created model using data gathered from the literature. The results of the simulations indicate that propagation in the transverse electromagnetic mode has the smallest attenuation and that for depths of up to 4000 m, there is an attenuation of less than 52 dB. Furthermore, the input impedance ranges from 10 Ω to 10 kΩ because of the uncertainties involved in the problem in question.

KEYWORDS

Electromagnetism; Oil Well; Wireless Telemetry; Monte-Carlo; Uncertainties; Heterogeneous Media

1. INTRODUCTION

Oil wells today are extremely complex and demand very expensive maintenance [1]. Modern drilling techniques can reach a few kilometres in depth, and the costs of a floating platform at open sea can cost billions of Brazilian Reais. Thus, it is critical to determine the internal conditions of the well, and indicators such as temperature, pressure and salinity must be constantly monitored [2].

Reliable telemetry by cables is very difficult to obtain due to the extreme conditions of the internal environment of an oil extraction well. As an example, there is continuous abrasion by sand and dirt that are carried by fluid flow. For this reason, the cables are periodically damaged and require replacements. Such replacements hinder the oil extraction process and, moreover, increase the costs due to the large amount of cables needed over the life of a single well.

The most obvious alternative to cabling is the adoption of wireless telemetry. However, the amount of electricity required to power wireless communication is not practical: power cables are needed because the use of batteries would lead to frequent maintenance, as they need to be recharged [1]. It is obvious, therefore, that the objective (and greater challenge) is to build a system without wires or batteries; that is, the downhole sensors must be powered and communicate with the base without the use of cables.

Several approaches are possible for wireless telemetry. One possibility would be the use of signal transmission techniques using a magnetic field [3]. In this method, a coil that is capable of inducing alternating current through the production pipe is used. This coil has sufficient intensity to transmit both power and the signal itself between the sensor and the base [4]. Through a second coil positioned at the sensor location, it is possible to recover the signal and power needed to feed and produce bilateral communication. However, the technique has a weakness: the production tube is not a transformer core, i.e., it is not designed to minimise the magnetic flux losses. Its hollow characteristic confers much loss by parasitic Foucault currents concomitantly with hysteresis losses, the cause of which is due to the inadequacy of the production tube material for magnetic purposes [5]. Therefore, for long distances, on the order of 3000 m or more, the method is unfeasible.

Another possible approach would be to analyse the oil well from the perspective of a coaxial propagation structure, which is formed by a conductor tube with a perfect centre (production tube) and a cylindrical shell that is coaxial with the tube and is also a perfect conductor. This approach makes it possible to conduct the analysis as it would be performed on a coaxial cable, and therefore, propagation occurs in the transverse electromagnetic mode [6]. This approach, however, has been studied without considering fluids of significant conductivity or possible uncertainties caused by temperature and electrical parameter variations of the fluid that fills the well.

The modelling difficulty concerns the fluid that fills the well. This fluid is heterogeneous because it exhibits significant variation in electric permittivity along the well depth. Furthermore, the actual fluid concentration varies from well to well, and at times within the well itself, which is another source of heterogeneity.

In this context, the present article proposes to study the electromagnetic propagation inside the annular area of oil wells. The goal is to develop a model that permits an approximation of the behaviour of the TEM propagation mode in the coaxial structure with losses and the previously cited heterogeneous characteristics. The first step is the deduction of the electric and magnetic field equations within the well, assuming a typical well structure. Thereafter, statistical models of the parameters that compose the medium are developed, considering their temperature variations and propagation frequency. Finally, the study uses Monte Carlo analysis to illustrate how the propagation attenuation behaves and to analyse the input impedance of the coaxial waveguide as a function of position.

2. MODELLING

2.1. Well Definition

The well will be modelled according to **Figure 1**. This figure depicts a coaxial guide bounded by perfect conductor metals with a homogeneous dielectric and losses. The upper extremity is completely closed by a perfect conductor metal to model the valve and metal duct assembly present in a wellhead. The bottom of the well is composed of concrete; however, this fact is neglected, and the well is assumed to behave as a semi-infinite waveguide.

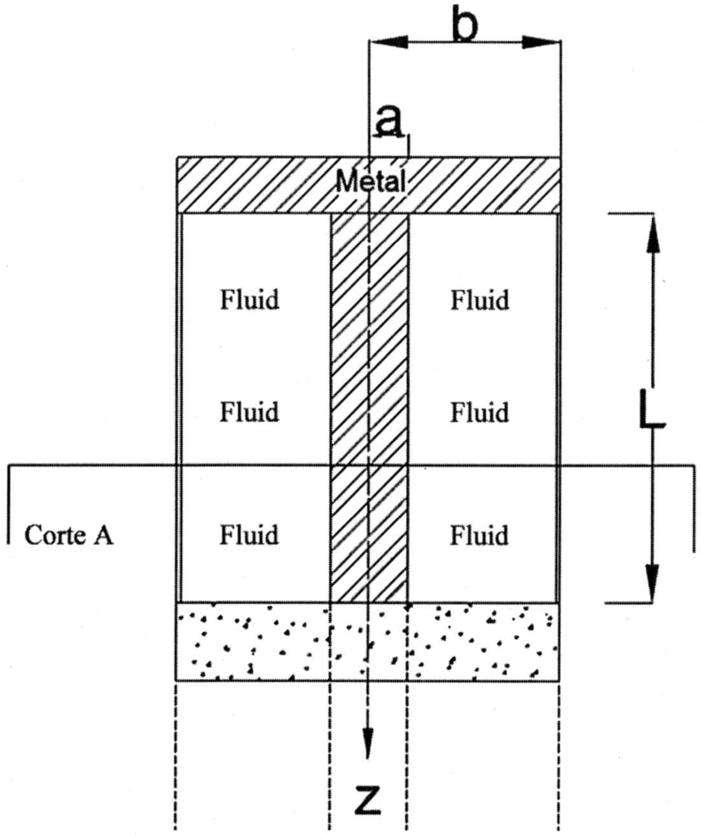

Figure 1. Adopted well structure. A semi-infinite well is assumed and is formed by two concentric cylinders and a metal block covering the top of the waveguide.

2.2. Deduction of the Propagation Equations

Modelling is performed using Maxwell's equations and the concept of vector potentials. The Maxwell's equations used are Equations (1) and (2):

$$\nabla \times \boldsymbol{E} = -\boldsymbol{M}_i - j\omega \boldsymbol{B} \tag{1}$$

$$\nabla \times \boldsymbol{H} = \boldsymbol{J}_i + \sigma_e \boldsymbol{E} + j\omega\varepsilon' \boldsymbol{E} \tag{2}$$

The real number constant ε' is the real part of the complex electrical permittivity of the medium. Furthermore, the complex constant σ_e is called effective conductivity and it represents the linear relation of conduction current and electric field on the medium.

Consider a medium with no source of magnetic charge $(\nabla \cdot \boldsymbol{B} = 0)$. Because $\nabla \cdot \nabla \times = 0$, the curl of A can be defined as

$$\nabla \times \boldsymbol{A} \equiv \boldsymbol{B} = \mu \boldsymbol{H} \tag{3}$$

Substituting 3 in 1 and considering $M_i = 0$ (Induced magnetic flux equal to zero), we obtain:

$$\nabla \times \boldsymbol{E} = -\nabla \times j\omega \boldsymbol{A} \tag{4}$$

Or:

$$\nabla \times (\boldsymbol{E} + j\omega \boldsymbol{A}) = 0 \tag{5}$$

Since $\nabla \times \nabla = 0$, an electric scalar potential g_e can be defined such that:

$$\boldsymbol{E} + j\omega \boldsymbol{A} \equiv -\nabla g_e \tag{6}$$

Applying the identity $\nabla \times \nabla \times = \nabla(\nabla \cdot) - \nabla^2$ in Equation (3) and using Equation (2), we obtain:

$$-\nabla\nabla\cdot\boldsymbol{A}+\nabla^2\boldsymbol{A}$$
$$=-\mu\boldsymbol{J}_i+\mu\left(\sigma_e+j\omega\varepsilon'\right)\left(\nabla g_e+j\omega\boldsymbol{A}\right)$$
$$\sigma_e\equiv\omega\varepsilon''+\sigma_s \tag{7}$$

In Equation (7), the parameter σ_e is the effective conductivity of the medium, while σ_s represents the static conductivity, that is, the conductivity when frequency is zero. At this moment, we are in a position to define $\nabla\cdot\boldsymbol{A}$. To simplify Equation (7), we use the gauge:

$$\nabla\cdot\boldsymbol{A}\equiv-\frac{\gamma^2}{j\omega}g_e \tag{8}$$

where $\gamma_1^2\equiv j\omega\mu\left(\sigma_e+j\omega\varepsilon'\right)$ is the complex constant of propagation of the medium. This step simplifies the equation for the electrical potential, resulting in:

$$\nabla^2\boldsymbol{A}=-\mu\boldsymbol{J}_i+\gamma^2\boldsymbol{A} \tag{9}$$

For homogeneous media, Equation (9), when solved, determines the electric vector potential in the medium, which can be used to obtain the equation for the electric field as a function of the electric vector potential:

$$\boldsymbol{E}=\frac{j\omega}{\gamma^2}\nabla\nabla\cdot\boldsymbol{A}-j\omega\boldsymbol{A} \tag{10}$$

We seek the solution in transverse electromagnetic mode (TEM), which is generated using $\boldsymbol{A}=\hat{a}_z A_z$ with the restriction $E_z = 0$, which leads to the following fields in cylindrical coordinates:

$$E_\rho=\frac{j\omega}{\gamma_1}\frac{\boldsymbol{A}}{\rho}\sinh\left(\gamma_1 z\right) \tag{11}$$

$$H_\phi = -\frac{1}{\mu}\frac{A}{\rho}\cosh(\gamma_1 z) \tag{12}$$

$$\gamma_1^2 \equiv j\omega\mu(\sigma_e + j\omega\varepsilon') \tag{13}$$

Note that because $E_\Phi = E_Z = 0$, the boundary conditions that require a null tangential component in the metal extremities are already met. As the well has a metal structure at one of the extremities, one must also ensure that E_ρ (z = 0) = 0. Note, however, that this condition is also already guaranteed.

The above solution represents a propagation model in TEM mode in a coaxial medium with losses inherent to the medium inside. However, it is important to remember that the solution was obtained by assuming a homogeneous propagation medium.

2.3. Statistical Modelling of the Medium Constituent Parameters

The propagation medium in the well consists of an oilbased dielectric fluid composed of water, oil and salts (typically $CaCl_2$) [7]. This fluid is the centre of all electromagnetic propagation, and therefore, its electrical characteristics must be examined in detail. A study over the range of 1 MHz to 100 MHz was previously conducted [7], revealing significantly variable behaviour based on the chemical composition of the fluid.

To model the variation of conductivity with frequency, the effective conductivity concept is applied [8] using the following formula:

$$\sigma_e \equiv \sigma_s + \omega\varepsilon'' \tag{14}$$

From the experimental curves obtained in [8], the parameters σ_s and ε'' can be calculated using a least squares method.

An analogous model can be created to model the variation in frequency of the real relative permittivity with the frequency. With

$$\varepsilon' \equiv \varepsilon_s + \kappa\omega \tag{15}$$

and using the experimental curves in [8], it is possible to estimate the parameters and κ using a least squares approach.

In addition to the variation in frequency, variation in the medium ε_s constituent parameter can also be observed in temperature. Such variation cannot be neglected because inside a 5000 m deep well, it is impossible to ensure that the temperature is uniform along its entire length. This fact, therefore, characterises non-homogeneity along the z direction.

To circumvent the situation, a model that assumes an average temperature in the medium is adopted. This temperature, in turn, is considered constant throughout the well, which implies a homogeneous medium. Thus, the influence of the variation of this average temperature on signal propagation in the well can be analysed, assuming a valid range of average temperatures.

Mathematically, a coefficient of correction in temperature is defined as the ratio between the temperature value in question and the value of the reference temperature, here defined as 25°C. Thus,

$$\theta \equiv \frac{\sigma_e(\theta)}{\sigma_e(25^\circ\text{C})} \tag{16}$$

$$\theta \approx a_1\theta^2 + a_2\theta + a_3 \tag{17}$$

$$\sigma_e = \sigma_e(25^{\circ}\mathrm{C}) \times \theta \tag{18}$$

The quadratic form was selected to present the best interpolating results using the experimental data from [8].

Similarly, there is increasing variation in the relative permittivity according to the temperature, as demonstrated by [7]. Again, it is used a quadratic model similar to the effective conductivity variation in temperature:

$$\theta_\varepsilon \equiv \frac{\varepsilon'(\theta)}{\varepsilon'(25^{\circ}\mathrm{C})} \tag{19}$$

$$\theta_\varepsilon \approx a_1\theta^2 + a_2\theta + a_3 \tag{20}$$

$$\varepsilon' = \varepsilon'(25^{\circ}\mathrm{C}) \times \theta_\varepsilon \tag{21}$$

3. EXPERIMENTS AND RESULTS

In all the experiments, a well with a length of 5000 m, an inner radius of 0.05 m and an outer radius of 0.1 m is assumed. All the Monte Carlo analyses were conducted with at least 1 million samples. The objective of the experiments was to obtain graphs and numerical values for the input impedance of the "oil well" waveguide and to analyse the propagation loss in the medium for 5000 m of depth.

All analyses were conducted over the range of 1 MHz to 100 MHz. For the other free parameters, the modelling used random variables whose distribution was selected to reflect their most common values. **Table 1** summarises the values selected for each free parameter of the previously described model and their respective distributions. For simplicity, when the random variable has simple and obvious domain restrictions, a uniform distribution was selected; otherwise, a normal distribution was used. Furthermore, statistical independence was assumed between the variables.

First, an experiment was conducted to calculate the attenuation in the well. The attenuation is directly dependent on the real part of the constant of propagation, whose square is defined as

$$\gamma_1^2 \equiv j\omega\mu\left(\sigma_e + j\omega\varepsilon'\right).$$

The models proposed in Section 2.3 were used for the conductivity and electric permittivity, and the relative magnetic permeability was assumed to be equal to 1.

The **Figure 2** presents a boxplot of the constant of attenuation for the various frequencies between 1 and 100 MHz. Values that appear in the graph as outliers are from highly unlikely combinations of the random variables of the problem and are most likely not physically feasible and, therefore, must be disregarded. The height of the boxes represents the range of values most likely to be observed, and its average point is a good approximation of the average. The boxplot of **Figure 3**demonstrates that the attenuation constant increases with an increase in frequency, as expected. Furthermore, note that the deviation of the coefficient increases with the elevation of the propagation frequency, which makes the system design even more difficult. Thus, it is clear that lower frequentcies are desirable from the point of view of signal attenuation.

Using the TEM mode equations presented in Equation (2), it is clear that the wave propagating in the direction of the bottom of the well has power attenuation given by $A = \exp\left(\alpha^2 L^2\right)$. Using the statistical analysis of the coefficient of attenuation, **Table 2** was generated.

Table 1. Random variables used for modelling uncertainties. For simplicity, when the random variable has simple and obvious domain restrictions, a uniform distribution was selected; otherwise, a normal distribution was used.

Variable	Average/a	Deviation/b	Distribution
σ_z (S/m)	0	4.0×10^{-5}	Uniform
ε_z (F/m)	3	12	Uniform
ε^{*} (F/m)	1.0×10^{-13}	3.0×10^{-12}	Uniform
k (F/H$_z$·m)	2.5×10^{-9}	5.0×10^{-10}	Normal
θ_{med} (°C)	40	5	Normal
a_1 (*permissivity*)	9.0×10^{-4}	2.0×10^{-4}	Normal
a_2 (*permissivity*)	-3.0×10^{-12}	6.0×10^{-3}	Normal
a_3 (*permissivity*)	1.2	0.1	Normal
a_1 (*permissivity*)	1.0×10^{-5}	2.0×10^{-6}	Normal
a_2 (*permissivity*)	0.0	5.0×10^{-4}	Normal
a_3(*permissivity*)	1.0	1.0×10^{-2}	Normal

The values from the table represent the upper limit of attenuation for 95% of the cases. Thus, for example, for a depth of 1000 m, P ($A \leq 13$ dB) = 0.95. The results presented in **Table 2** do not agree with the propagation for L = 5000 m of depth because of the high-energy attenuation (130.3 dB). However, it is essential to note that the table represents an upper limit of attenuation, taking 95% of the possible combinations of propagation medium and an average temperature. Moreover, as the table demonstrates, propagations up to 2000 m depth are highly acceptable, as an attenuation of up to 52.1 dB is observed in current communication technology. Even at greater depths, the possibility of communication cannot be excluded.

The second experiment was performed to analyse the input impedance of the coaxial waveguide as a function of the excitation source position. The experiment was conducted at 1 MHz.

Setting the impedance at any point as the ratio E_ρ / H_ϕ, we obtain

$$Z = Z_0 \tanh(\gamma_1 z) \tag{22}$$

$$-L < z \leq 0 \tag{23}$$

$$Z_0 = -\mu \frac{j\omega}{\gamma_1} \tag{24}$$

The position selected for analysis was at 1/4 wavelength from the wellhead, that is, $z = \lambda/4$. However, the wavelength itself is a random variable because $\lambda = 2 * \pi / \beta$ and $\beta = \mathrm{Im}\{\gamma\}$ is a random variable. Monte Carlo analysis revealed that λ varies between 90 m and 170 m for 90% of the cases, leading to the selection of position $Z = ((90+170)*1/2)/4 = 32.5$ m.

Using Monte Carlo analysis again, it can be observed in **Figure 4** that the input resistance (real part of the impedance) of the system exhibits great variation, ranging from 10 Ω to 1.0 kΩ for 90% of the cases. This variation is due to the variation of the relative permittivity in the medium. Therefore, it is necessary to design a generator circuit that provides a good match for a wide range of input impedances, i.e., the generator/receiver circuit must lose as little power as possible by reflection.

Table 2. Estimated attenuation of electromagnectic waves propagating in the annular region of the well.

L (m)	Electric field attenuation (dB)	Power attenuation (dB)
1000	13.0	26.1
2000	26.1	52.1
3000	39.1	78.2
4000	52.1	104.2
5000	65.1	130.3

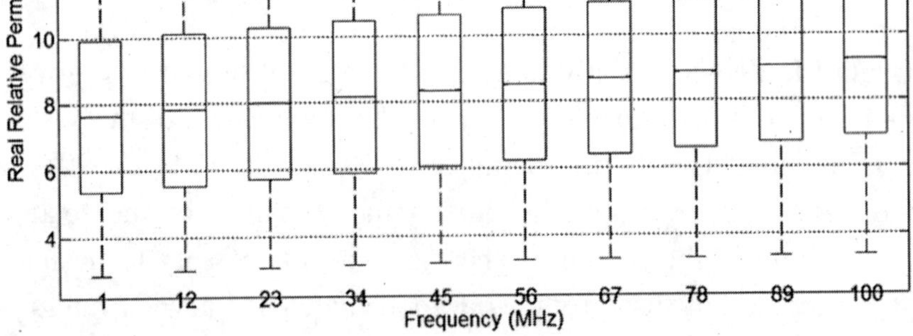

Figure 2. Boxplot of relative permittivity as a function of frequency. The horizontal axis represent the frequency of propagation in MHz, while the vertical axis represent the actual value of the electrical permittivity.

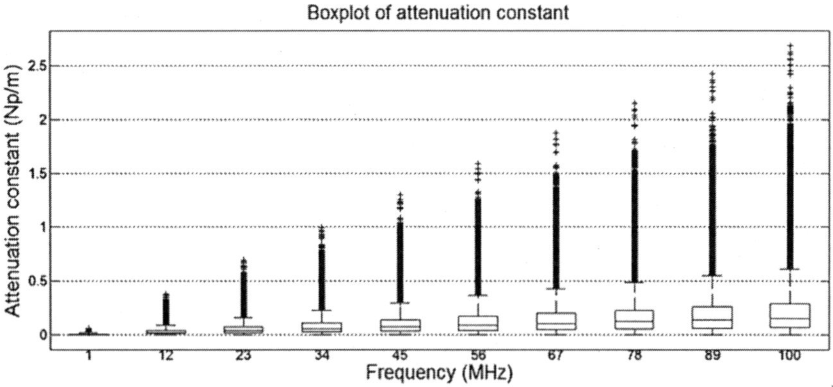

Figure 3. Boxplot of attenuation constant as a function of frequency.

4. CONCLUSIONS

The present study focused on analysis of the electromagnetic propagation inside the annular area of oil wells, assuming the interior medium is composed of dielectrics with significant conductivity. The well behaviour was evaluated with respect to the input impedance and propagation attenuation.

To quantify the well behaviour with respect to electromagnetic propagation, a coaxial waveguide model was developed, modelling the wellhead as a metal cap that completely closes one extremity of the waveguide. First, we solved the wave equation for a homogeneous cylindrical coaxial waveguide, resulting in an analytical model. Then, an uncertainty model was adopted to approximate the heterogeneous and imprecise characteristics of the fluid that functions as a dielectric inside the well.

Using this model, two experiments were developed, both by simulation. The first aimed at analysing the input impedance observed by a source positioned at 1/4 of the wavelength from the wellhead, while the other aimed at analysing the attenuation as a function of length or well depth.

Due to the uncertainties present, the input impedance was observed to vary from 10 Ω to 1 kΩ for the frequency of 1 MHz. The conclusion was obtained by Monte Carlo simulation and was applied to the expression derived for the waveguide input impedance.

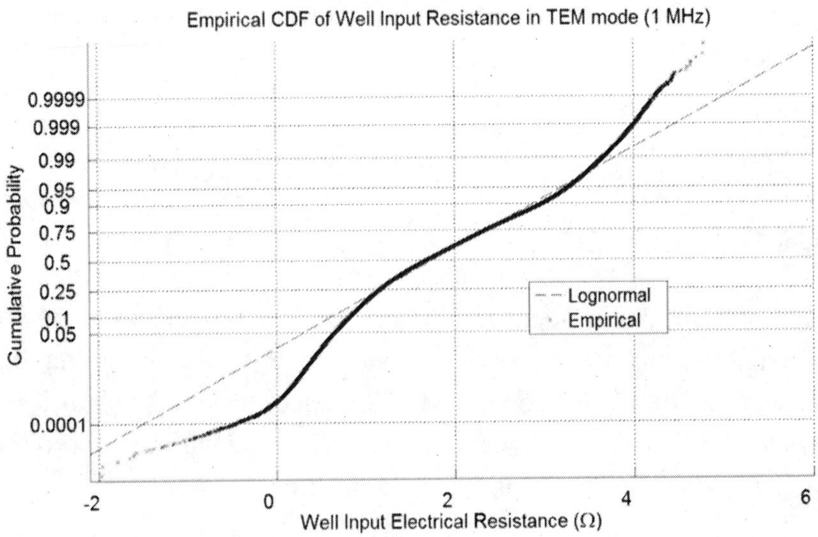

Figure 4. Empirical cumulative distribution function of the well input resistance in TEM mode at 1/4 wavelength from well head. The horizontal axis show the log on base 10 of the value of input resistance in Ohms, while the vertical axis show the cumulative distribution function value.

By performing the attenuation analysis, it was concluded that for 95% of the cases, the constant of attenuation in TEM mode is less than 0.8×10^{-1} Np/m for a frequency of 1 MHz. The power attenuation at 4000 m depth was also observed to be approximately 100 dB for the same frequency of 1 MHz, with 95% probability. Although 100 dB appears to be a large attenuation, most wells in operation are between 1000 m and 2000 m in length, and the attenuation is much lower at these depths.

REFERENCES

1. S. Brilles, "Remote Downhole Well Telemetry," US Patent US6766141, 2004.

2. J. A. D. Rosa, A. J. Carvalho and R. D. S. Xavier, "Petroleum Reservoir Engineering," Rio de Janeiro, 2006.

3. F. Sakata, H. Wakiwaka, M. Hanabusa, N. Yamazaki and H. Yamada, "Performance Analysis of Long Distance Transmitting of a Magnetic Signal in a Cylindrical Steel Rod," IEEE Translation Journal on Magnectics in Japan, Vol. 8, No. 2, 1993, pp. 102-106.

4. F. Harold J. Vinegar, R. R. Burnett, G. C. W. M. Savage and J. W. Hall, "Permanent Downhole, Wireless, TwoWay Telemetry Backbone Using Redundant Repeaters," US Patent US6633236B2, 2003.

5. B. W. Kennedy, "Energy Efficient Transformers," New York, 1998.

6. K. A. Safynia and R. W. McBride, "System and Method for Communicating Signals in a Cased Borehole with Tubing," US Patent US4839644, 1989.

7. P. A. Patil, et al., "Experimental Study of Electrical Properties of Oil-Based Mud in the Frequency Range from 1 to 100

MHz," SPE Drilling and Completion, Vol. 25, No. 3, 2010, pp. 380-390.

8. C. A. Balanis, "Advanced Engineering Electromagnetics, Vol. 52, No. 1," Wiley, Hoboken, 1989, p. 1008.

CHAPTER 8

Analytical Solution for Acoustic Waves Propagation in Fluids

Mohamed I. A. Othman[1], Mohamed Galal Sayed Ali[2], Roushdi Mohamed Farouk[3]

[1,3] Department of Mathematics, Faculty of Science, Zagazig University, Zagazig, Egypt

[2] Department of Physics, Faculty of Science, Minia University, Minia, Egypt

ABSTRACT

This paper presents a mathematical model of linear acoustic wave propagation in fluids. The benefits of a mathematical model over a normal mode analysis are first discussed, then the mathematical model for acoustic propagation in the test medium is developed using computer simulations. The approach is based on a analytical solution to the homogeneous wave equation for fluid medium. A good agreement between the computational presented results with published data.

KEYWORDS

Acoustic waves, normal mode analysis, fluid medium

1. INTRODUCTION

In recent years, physical acoustic wave modeling has become a successful tool in diagnostic and therapeutic ultrasound application. There are several wave equations available for describing acoustic wave propagation [1-4]. Numerical methods can be used as a tool for sound field simulation. Discrete-time simulation algorithms for wave propagation can be derived by numerically solving a acoustic wave equation in terms of the variables for sound pressure and particle velocity. Initial conditions for time derivatives and boundary conditions for space derivatives are necessary to provide a complete set of solutions of the wave equation. These equations are most commonly solved by propagation in time. However, when propagating over large distances, such methods are expensive in terms of memory and computational costs [5].

The normal mode method analysis gives exact solutions without any assumed restrictions on pressure and velocity components distributions. It is applied to wide range of problems in different branches (Othman [6-8], Sharma et al. [9], Othman and Kumar [10], Othman and Singh [11] and Othman et al. [12]). It can be applied to boundary-layer problems, which are described by the linearized Navier-stokes equations in electrohydrodynamic (Othman [13]).

In this paper, the normal mode analysis can be employed to solve linear acoustic wave equation analytically. The technique focuses on description of a linear model and discuses the conditions under which using this technique. The propagation of acoustic pressure wave by the normal mood analysis in a medium with two-dimensional spatially-variable acoustic properties has been explained.

2. ACOUSTIC WAVE EQUATION

Consider sound waves propagating in the water. Instead of the wave equation, we base our work on the basic Euler's equation and the equation of continuity. For simplicity, the discussion is confined to a two-dimensional space. In a 2-D Cartesian coordinate system, the sound pressure p and the particle velocity v satisfy the following linear equations:

$$-\rho\nabla\cdot\mathbf{v}(x,y,t)=\frac{1}{c^2}\frac{\partial p(x,y,t)}{\partial t} \tag{1}$$

$$\frac{\partial p(x,y,t)}{\partial x}=-\rho\frac{\partial v_x(x,y,t)}{\partial t}, \tag{2}$$

$$\frac{\partial p(x,y,t)}{\partial y}=-\rho\frac{\partial v_y(x,y,t)}{\partial t} \tag{3}$$

where $\mathbf{v}(x,y,t)=\hat{x}v_x(x,y,t)+\hat{y}v_y(x,y,t)$ is the particle velocity, p(x, y, t) is the pressure and ρ is the density of the fluid with wave number $k=\frac{\omega}{c}+i\alpha$ where $i=\sqrt{-1}$, ω is the angular frequency, c and α are the speed of sound and attenuation in inhomogeneous medium, respectively.

3. NORMAL MODE ANALYSIS

The solution of considered physical variable can be decomposed in terms of normal modes as the following form

$$\left[v_x, v_y, p\right](x, y, t) = \left[v_x^*, v_y^*, p^*\right](x) e^{i(\omega t + ky)}, \tag{4}$$

where v_x^*, v_y^* and p^* are the amplitude of the functions v_x, v_y and p respectively.

Equations (1)-(3) become

$$-\rho D v_x^* - i\rho k v_y^* = \frac{i\omega}{c^2} p^*, \tag{5}$$

$$D p^* = -i\rho\omega v_x^*, \tag{6}$$

$$k p^* = -\rho\omega v_y^*. \tag{7}$$

where, $D = \frac{\mathrm{d}}{\mathrm{d}x}$.

Equations (5)-(7) form a coupled system Eliminating v_x^* and v_y^* between Equations (5)-(7) we obtain

$$D^2 p^* - m^2 p^* = 0. \tag{8}$$

where, $m^2 = \left[k^2 - \frac{\omega^2}{c^2}\right]$.

The solution of Equation (8) has the form

$$p^*(x) = \sum_{i=1}^{2} p_i^*. \tag{9}$$

where $p_i^*(x) = G_i(k,\omega)e^{r_i x}$, and r_i are the roots of the characteristic equation

$$r_i^2 - m^2 = 0 \tag{10}$$

The solution of Equation (8) is given by

$$p^*(x) = G_1(k,\omega)e^{-mx} + G_2(k,\omega)e^{mx} \tag{11}$$

From Equations (6) and (11) we can obtain

$$v_x^*(x) = G_1'(k,\omega)e^{-mx} + G_2'(k,\omega)e^{mx} \tag{12}$$

where,

$$G_1'(k,\omega) = \frac{m}{i\rho\omega}G_1(k,\omega),\ G_2'(k,\omega) = -\frac{m}{i\rho\omega}G_2(k,\omega). \tag{13}$$

From Equations (7) and (11) we can obtain

$$v_y^*(x) = G_1''(k,\omega)e^{-mx} + G_2''(k,\omega)e^{mx} \tag{14}$$

where,

$$G_1''(k,\omega) = -\frac{k}{\rho\omega}G_1(k,\omega),\ G_2''(k,\omega) = -\frac{k}{\rho\omega}G_2(k,\omega). \tag{15}$$

4. BOUNDARY CONDITIONS

On the surface at x = 0

$$v_x^* = F(y,t),\ p^* = p_0(y,t). \tag{16}$$

Substituting from (4) into (16) then Equations (11) and (14)

$$p_0^* = G_1 + G_2 \tag{17}$$

$$\frac{i\rho\omega}{m} F_0^* = G_1 - G_2, \tag{18}$$

By adding Equations (17) and (18) we obtain

$$G_1 = \frac{1}{2}\left(P_0^* + \frac{i\rho\omega}{m} F_0^* \right) \tag{19}$$

By subtracting Equations (17) and (18) we get

$$G_2 = \frac{1}{2}\left(P_0^* - \frac{i\rho\omega}{m} F_0^* \right) \tag{20}$$

By substituting from Equations (19) and (20) into Equations (11), (12) and (14)

$$P(x, y, t) = \left[G_1 e^{-mx} + G_2 e^{mx} \right] e^{i(\omega t + ky)} \tag{21}$$

$$v_x(x, y, t) = \frac{m}{\rho\omega}\left[G_1(k, \omega) e^{-mx} - G_2(k, \omega) e^{mx} \right] e^{i((\omega t + ky)} \tag{22}$$

$$v_y(x, y, t) = -\frac{k}{\rho\omega}\left[G_1(k, \omega) e^{-mx} + G_2(k, \omega) e^{mx} \right] e^{i(\omega t + ky)} \tag{23}$$

5. COMPUTATIONAL RESULTS

To study the wave propagation phenomenon in viscous medium and under different frequencies, we can apply the theoretical acoustic viscous wave Equation (21). Using water as the medium, the parameters are given as following: $\rho = 998 \text{ Kg/m}^3$ and $c = 1481 \text{ m/sec}$. Let the wave peak amplitude be $P_o^* = 1$ Pa and $F_o^* = 1$ m/sec at the source (x = 0), we simulate the pressure wave peak amplitude, in Equation (21), vs. the distance from the source at various frequencies. The results are shown in **Figure 1** (in dB and linear scale). As expected, the peak wave amplitude becomes smaller as we move further from the source. We can also notice that as the wave frequency goes higher, more attenuation can be observed at any given location. The peak pressure as function of frequency shown in **Figure 2** for fixed distance at x = 2 cm. It is clear from this figure that the magnitude of peak of pressure little changes with the frequency. The predicted results are very agreement to the recorded by Wang [14].

Let us consider a 2-D simulation in which the pressure varies in the x and y directions. **Figure 3**shows the 2-D pressure computational as function in the plane x-y. From this figure, it can be seen that the pressure amplitude becomes smaller when moving in the x-y plane further from the source.

6. CONCLUSIONS

A normal mode analysis which accurately the pressure acoustic wave equation, has been developed. This technique has a number of attractive features, foremost of which is the speed and simplicity with which it can be designed and implemented. The model could be used in the future to incorporate non-linear propagation effects.

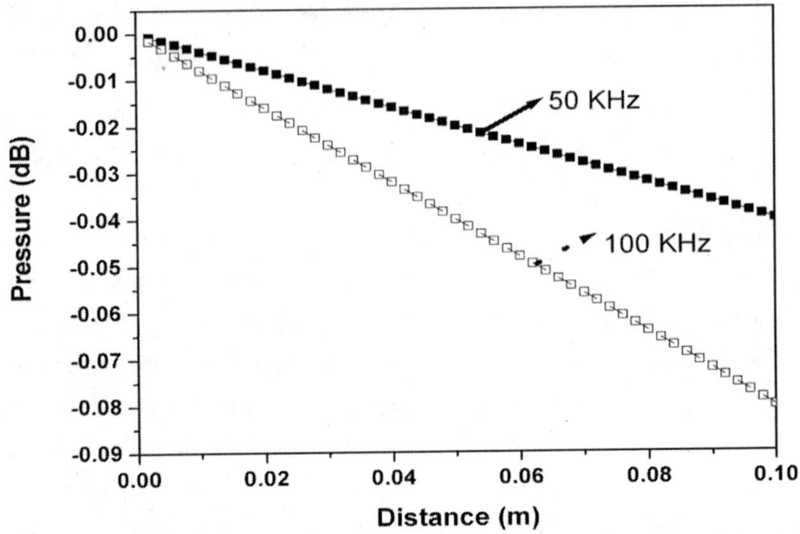

Figure 1. Pressure amplitude (dB) as function of the distance from the origin.

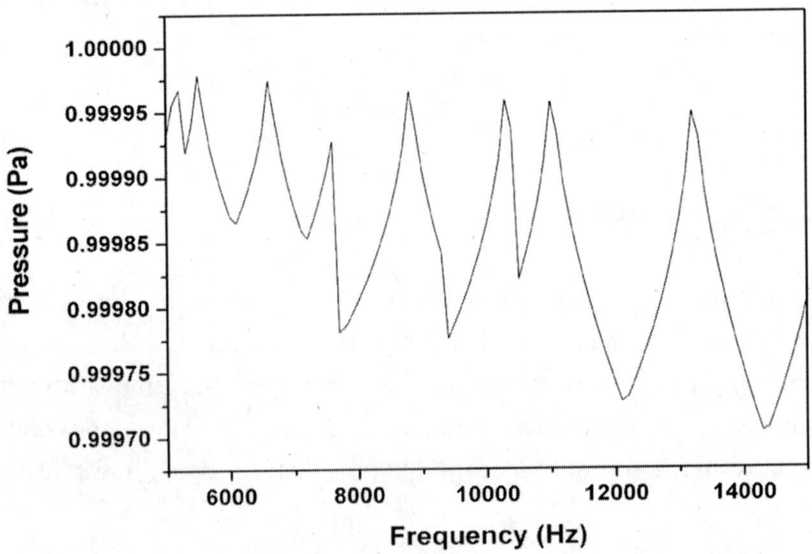

Figure 2. The pressure amplitude as function of frequency at distance 2 cm from the source.

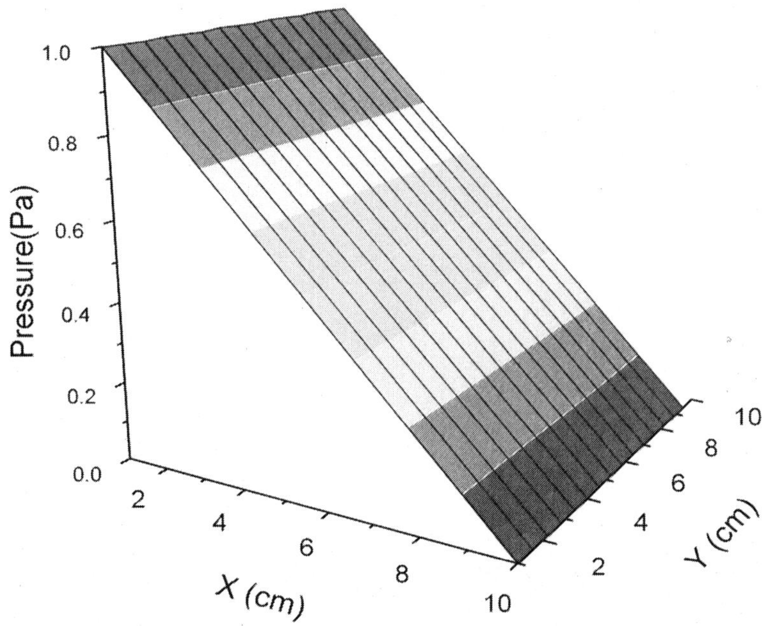

Figure 3. The pressure amplitude as function of plane x-y direction.

REFERENCES

1. L. E. Kinsler, A. R. Frey, A. B. Coppens and J. V. Sanders, "Fundamentals of Acoustics," 4th Edition, Wiley, New York, 2000.

2. S. W. Rienstra and A. Hirschberg, "An Introduction to Acoustics," Eindhoven University of Technology, Eindhoven, 2009.

3. S. W. Rienstra, "Sound Propagation in Slowly Varying Lined Flow Ducts of Arbitrary Cross Section," Journal of Fluid Mechanics, Vol. 495, 2003, pp. 157-173.

4. D. Daniel and Joseph, "Review Potential Flow of Viscous Fluids: Historical Notes," International Journal of Multiphase Flow, Vol. 32, No. 3, 2006, pp. 285-310.

5. J. G. Maloney and K. E. Cummings, "Adaptation of FDTD Techniques to Acoustic Modeling," 11th Annual Review of Progress in Applied Computational Electromagnetics, Vol. 2, 1995, pp. 724-731.

6. M. I. A. Othman, "Effect of Rotation in Case of 2-D Problems of Generalized Thermoelasticity with Thermal Relaxation," Mechanics & Mechanical Engineering, Vol. 8, 2005, pp. 111-126.

7. M. I. A. Othman, "Effect of Rotation on Plane Waves in Generalized Thermo-Elasticity with Two Relaxation Times," International Journal of Solids and Structures, Vol. 41, No. 11-12, 2004, pp. 2939-2956.

8. M. I. A. Othman, "Lord-Shulman Theory under the Dependence of the Modulus of Elasticity on the Reference Temperature in Two-Dimensional Generalized Thermo-elasticity," Journal of Thermal Stresses, Vol. 25, No. 11, 2002, pp. 1027-1045.

9. J. N. Sharma, R. Chand and M. I. A. Othman, "On the Propagation of Lamb Waves in Visco-Thermoelastic Plates under Fluid Loadings," International Journal of Engineering Science, Vol. 47, No. 3, 2009, pp. 391-404.

10. M. I. A. Othman and R. Kumar, "Reflection of Magneto-Thermoelastic Waves under the Effect of Temperature Dependent Properties in Generalized Thermo-Elasticity with Four Theories," International Communications in Heat and Mass Transfer, Vol. 36, No. 5, 2009, pp. 513-520.

11. M. I. A. Othman and B. Singh, "The Effect of Rotation on Generalized Micropolar Thermoelasticity for a Halfspace under Five Theories," International Journal of Solids and Structures, Vol. 44, No. 9, 2007, pp. 2748- 2762.

12. M. I. A. Othman, Kh. Lotfy and R. M. Farouk, "Generalized Thermo-Micro-Stretch Elastic Medium with Temperature Dependent Properties for Different Theories," Engineering Analysis of Boundary Element, Vol. 34, No. 3, 2010, pp. 229-237.

13. M. I. A. Othman, "Electrohydrodynamic Stability in a Horizontal Viscoelastic Fluid Layer in the Presence of a Vertical Temperature Gradient," International Journal of Engineering Science, Vol. 39, No. 11, 2001, pp. 1217- 1232.

14. S. Wang, "Finite-Difference Time-Domain Approach to Underwater Acoustic Scattering problems," Journal of the Acoustical Society of America, Vol. 99, No. 4, 1996, pp. 1924-1931.

CHAPTER 9

Antiplane elastic wave propagation in pre-stressed periodic structures; tuning, band gap switching and invariance

Ellis G. Barnwell, William J. Parnell, I. David Abrahams

HIGHLIGHTS

- Wave propagation in a pre-stressed periodic structure is considered.
- Band diagrams are found using the theory of small-on-large and it is shown that they are modified by pre-stress.
- It is found that stop bands can be switched on or off by application of a pre-stress.
- The choice of material has a significant effect on the stop band structure.
- By studying the notion of invariance, the concept of a phononic cloak is introduced.

ABSTRACT

The effect of nonlinear elastic pre-stress on antiplane elastic wave propagation in a two-dimensional periodic structure is investigated. The medium consists of cylindrical annuli embedded on a periodic square lattice in a uniform host material. An identical inhomogeneous deformation is imposed in each annulus and the theory of small-on-large is used to find the incremental wave equation governing subsequent small-amplitude antiplane waves. The plane-wave-expansion method is employed in order to determine the permissable eigenfrequencies. It is found that pre-stress significantly affects the band gap structure for Mooney–Rivlin and Fung type materials, allowing stop bands to be switched on and off. However, it is also shown that for a specific class of materials, their phononic properties remain invariant under nonlinear deformation, permitting some rather interesting behaviour and leading to the possibility of *phononic cloaks*.

KEYWORDS

Phononic switch; Configurable; Tunable; Band gaps; Pre-stress; Hyperelastic

1. INTRODUCTION

Complex materials are at the heart of modern-day engineering and technology. A vast number of such media possess an intricate microstructure that offers macroscopic behaviour that is not present in naturally occurring materials. In the field of acoustics and elastodynamics, phononic media and metamaterials present a wide array of opportunities for supporting and directing a multitude of wave types [1] and [2]. The ability to design these materials for a variety of wave filtering, focusing and directivity applications is greatly advantageous. In particular, it is desirable to create macroscopic properties that can be tuned in real time (so called *configurable* or *re-configurable*phononic media), rather than the medium having fixed dynamic properties from the outset. Predicting how the microstructure affects the macroscopic response on a static and dynamic level is pivotal to such design and control of these materials.

Employing inhomogeneous media in order to allow or restrict wave propagation has been discussed by numerous authors over the last century. The analysis of band gap formation in periodic elastic and acoustic media took off in earnest in the early 1990s, see e.g. [3],[4], [5] and [6] and the excellent recent review paper [7] for a broader literature review. Early work in tuning the acoustic response of such media focused on modifying band structures by rotating cylinders [8], [9] and [10]. Others have utilised the tunability of a phononic crystal to create wave guides [11], [12] and [13]. These latter works, however, generally rely on tuning the material at the design and construction phase, rather than allowing real-time tuning post-construction. This applies similarly to the work in photonics by e.g. Zhang et al. [14], which described how photonic band gaps could be enlarged by the use of a small volume fraction of a third material phase. In these situations, once manufactured, the medium has a fixed dynamic range.

As a method of performing external control over a crystal, several research groups have considered the influence of magnetic or electric fields [15] and [16]. In [17], an electric field is used to deform annular cylinders which are embedded in air. The use of piezoelectric materials has also been studied [18], [19] and [20] and similarly, electro-rheological materials have recently been utilised [21]. Temperature control has been suggested as a mechanism for tuning [22] and [23] although typically the temperature effect on elastic properties is so small that control is only achievable through a phase transition. This mechanism provides the advantage of offering a phononic switch in such materials. Mechanical mechanisms to create *photonic* switches have also recently been considered, see e.g. [24] and [25].

The effect of a nonlinear pre-stress on elastic wave propagation has been studied previously in one dimension [26] and [27]. In [27] the stop and pass band structure was investigated and it was shown that the band structure can be manipulated effectively using pre-stress. Similar work was carried out in which an electric bias was studied as a mechanism for controlling the deformation in layered composites [28]. Another example in one dimension is the extension of an inhomogeneous beam [29], in which the manipulation and annihilation of stop bands is studied. An experimental investigation into one-dimensional structures was carried out in [30]. By making use of numerical simulations, via the finite-element method, the mechanical tunability of three-dimensional structures has also recently been studied [31]. Finally, the effect of microstructural buckling of an elastic material was the focus of investigation in [32] and [33].

The objective of this article is to introduce a new mechanism for tuning band gaps in the context of horizontally polarised shear (SH or antiplane) elastic waves, enabling the design of reversible phononic switches and configurable phononic media via a novel material geometry. The phononic medium studied is a two-dimensional periodic array of incompressible nonlinear elastic annular cylinders embedded in a homogeneous elastic host as depicted in Fig. 1. SH waves are polarised along the axes of the cylinders and propagate in the plane orthogonal to these axes. As shall be described in Section 2, the annular regions are pre-stressed whilst the host medium remains stress free. The influence of this pre-stress on linear wave propagation through the two-dimensional medium is then studied by first deriving the governing wave equation via application of the theory of small-on-large. In

Section 3 the plane-wave-expansion (PWE) method is employed (in modified form) and extended in order to take into account the fact that the waves are propagating through a pre-stressed structure where the incremental moduli are strongly inhomogeneous. This method determines the associated dispersion curves very efficiently. In Section 4 the effect of the pre-stress on the structure is determined. Some concepts associated with phononic invariance are introduced including phononic equivalence and phononic cloaks. The possibility of phononic switching is then investigated for a variety of choices of strain energy function in the annulus phase. This new approach also permits the study of inhomogeneous deformation over and above changes in material geometry alone. Concluding remarks are offered in Section 5.

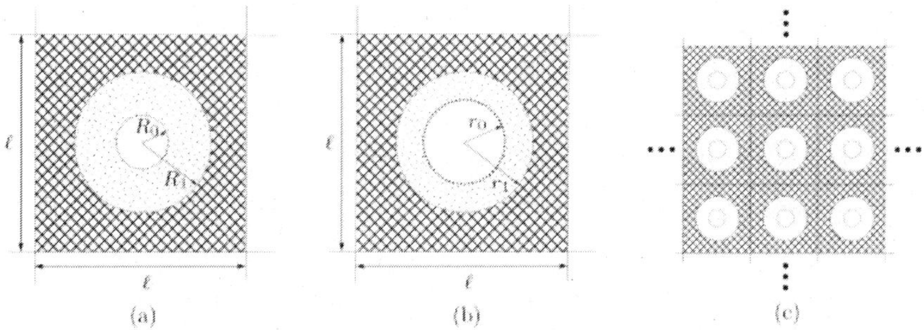

Figure. 1. Illustration of the unit cell for (a) the undeformed configuration and (b) the deformed configuration. The cylindrical annuli are embedded in a square periodic lattice with period ℓ in a stress-free, homogeneous medium shown by the hashed pattern. The periodic arrangement is illustrated in (c).

The restriction to the prescribed deformation, and to SH waves, is for clarity of exposition, focusing on the effect of the nonlinear pre-stress on wave propagation and the development of the new plane-wave-expansion technique. It also enables rather interesting results to be derived associated with invariance of phononic properties to pre-stress. More complex deformations and structures, as well as the full two-dimensional in-plane coupled compressional and shear wave problem will be described elsewhere.

2. CYLINDER INFLATION AND THE INCREMENTAL WAVE EQUATION

In this section the geometry and properties of the medium are first described together with the initial deformation and the incremental wave equation governing superposed antiplane waves inside the material. As illustrated in Fig. 1 the medium consists of a two-dimensional square array (with period ℓ) of nonlinear elastic annular cylinders each with initial inner and outer radii R_0 and R_1 respectively, density ρ_1 and linear elastic shear moduli μ_1 embedded inside a homogeneous elastic host medium of density ρ_0 and linear elastic shear modulus μ_0. The unit cell is depicted in its stress-free state inFig. 1(a). Each cylindrical annulus can in principal be pre-stressed independently but, in order to maintain periodicity, all are assumed to be subject to the same deformation. The latter consists of an inflation (leading to an inhomogeneous radial deformation) and extension along its axis such that, crucially here, the outer radius of the cylinder remains unchanged, $r_1=R_1$. This results in a deformed configuration as depicted in Fig. 1(b). The host medium may be linear or nonlinear elastic. Since it is not subject to initial deformation this is irrelevant in the context of the problem studied here.

In order to determine the manner by which small amplitude waves propagate in the pre-stressed solid, the theory of small-on-large is employed. The majority of the technical details will be omitted here but can be found elsewhere, e.g. in [34] and [35]. In essence, the theory is based on a linearisation about the pre-stressed state, determining the equations that govern incremental waves in a body that has exhibited a finite static deformation.

It is assumed that the interior of each annular cylinder is filled with an inviscid gas such that no shear waves can propagate inside this cavity. This means that the SH waves are subject to traction-free conditions on the cavity boundary since no antiplane traction is imposed by the pre-stress. The precise deformation used in this paper was previously utilised for generating infinite [36] and finite [37] hyperelastic cloaks.

Although the deformation studied here may be considered as somewhat idealised, it exhibits similar features to other deformation states associated with more complex geometries. Furthermore, since the deformation can be defined in simple mathematical (analytical) form, the incremental moduli that arise in the governing wave equation can be written explicitly. This makes the modified application of the plane-wave-expansion method described in Section 3 more straightforward in this initial step. More complex lattice geometries could also be studied if one wished to incorporate additional tunable characteristics such as material anisotropy or multi-phase microstructure.

2.1. Initial deformation

The initial deformation of the annular region is the same as that used in [37] and [38]. The equations of nonlinear elasticity are considered with boundary conditions imposed on the inner and outer faces of the annulus and an imposed axial stretch. Body forces are neglected.

Using a cylindrical polar coordinate system with its origin at the centre of a cell, referred to as the fundamental periodic cell, consider an inflation of the cylindrical annulus due to an imposed pressure, p_0, acting on its inner surface, accompanied by an axial stretch ζ. The external face of the annulus is traction free. The deformation that ensues is written formally as

$$R = R(r), \qquad \Theta = \theta, \qquad Z = \frac{z}{\zeta}, \tag{1}$$

for some function R(r) to be determined. The usual convention is followed, using upper (lower) case variables for the reference (deformed) configuration. The principal stretches in the directions r,θ and z respectively are given by

$$\lambda_r = \frac{1}{R'(r)}, \qquad \lambda_\theta = \frac{r}{R(r)}, \qquad \lambda_z = \zeta, \tag{2}$$

where the prime denotes differentiation with respect to the argument. The inner radius is permitted to change whilst the outer radius is fixed at R_1 by choosing ζappropriately for the imposed pressure, i.e.

$$R_0 = R(r_0), \qquad R_1 = R(r_1) = r_1. \tag{3}$$

The second of (3) together with incompressibility ($\lambda_r\lambda_\theta\lambda_z=1$) gives

$$R(r) = \sqrt{\zeta(r^2 + M)}, \tag{4}$$

where $M = (1/\zeta - 1)R_1^2$. One can either choose ζ and use (4) to find r_0, or choose $r_0$and rearrange (4) to find ζ.

The radial Cauchy stress induced in the annular cylinders by the initial deformation can be determined by integrating the radial equation of equilibrium so that [39]

$$T_{rr}(r) = -p_0 + \int_{r_0}^{r} \frac{1}{s}\left(\lambda_\theta(s)\frac{\partial W}{\partial \lambda_\theta} - \lambda_r(s)\frac{\partial W}{\partial \lambda_r}\right) ds, \tag{5}$$

where W is the strain energy of the hyperelastic cylinder and $T_{rr}(r_0)=-p_0$ has been imposed on the inner surface. It is further stressed also that $T_{rr}=0$ on $r=r_1$.

The aim now is to understand the influence that this deformation (within each annular cylinder) has on small-amplitude SH-wave propagation through the structure. It should be noted that the tractions required to apply the desired change in inner radius described above are taken to act only on the annular region and are maintained whilst the waves propagate. By ensuring that the outer radius remains fixed, no force is applied to the host phase and therefore it remains undeformed. One can consider a number of mechanisms to ensure that this deformation is possible in practice, including the consideration that the interface between the annulus and host allows sliding on a long timescale corresponding to the application of the finite deformation, whilst not permitting sliding on the short timescale associated with the small-amplitude linear elastic waves.

2.2. Incremental wave equation

Assume time-harmonic motion of the form $\exp(-i\omega t)$ where ω is the angular frequency of the wave. As with the initial deformation, the incremental wave equation in the periodic cell can be described effectively by using the polar coordinate system. However, when looking for global solutions, a slightly different approach is required as shall be seen shortly.

The theory of small-on-large can be employed to find the wave equation in the deformed annular region of the fundamental periodic cell. This theory is essentially a linearisation about deformed state, describing small-amplitude waves in the pre-stressed medium. The equation governing incremental antiplane waves polarised in the z direction, with displacements $w=w(r,\theta)$ is given by [36], [37] and [38]

$$\frac{1}{r}\frac{\partial}{\partial r}\left(r\mu_r(r)\frac{\partial w}{\partial r}\right)+\frac{1}{r^2}\frac{\partial}{\partial\theta}\left(\mu_\theta(r)\frac{\partial w}{\partial\theta}\right)+\rho\,\omega^2 w=0, \tag{6}$$

where $\rho=\rho_1$ is the constant mass density in the annulus $r_0<r<r_1$. Note that since the material is incompressible the density of the annulus is unaffected by the pre-stress. As the limit of no deformation is approached, the equation above reduces to the usual linear wave equation.

The incremental shear moduli $\mu_r(r)$ and $\mu_\theta(r)$ are dependent on the specific form of strain energy function, $W(\lambda_r,\lambda_\theta,\lambda_z)$, that is used to model the nonlinear elastic response of the annular cylinders, and their general forms can be written down as [35] and [38]

$$\mu_r(r)=\left(\frac{\lambda_r\frac{\partial W}{\partial\lambda_r}-\lambda_z\frac{\partial W}{\partial\lambda_z}}{\lambda_r^2-\lambda_z^2}\right)\lambda_r^2,\qquad \mu_\theta(r)=\left(\frac{\lambda_\theta\frac{\partial W}{\partial\lambda_\theta}-\lambda_z\frac{\partial W}{\partial\lambda_z}}{\lambda_\theta^2-\lambda_z^2}\right)\lambda_\theta^2. \tag{7}$$

Note that the shear moduli in the pre-stressed annular cylinders are inhomogeneous and exhibit cylindrical anisotropy due to the imposed deformation and induced stress fields. When there is no pre-stress $\mu_r=\mu_\theta=\mu_1$. Because the host medium remains stress free, the shear moduli in this region are isotropic and homogeneous. The governing wave equation in the host, $r>r_1$, is then given by (6) with $\mu_r=\mu_\theta=\mu_0$ and$\rho=\rho_0$. The region inside the annular cylinder is occupied by an inviscid fluid at constant pressure and hence no shear waves can propagate in this region.

As described above, the incremental equation (6) applies on the fundamental periodic cell, i.e. to each annular cylinder with its centre as the origin of a local cylindrical polar coordinate system. However, solving this equation globally for the wave and imposing appropriate periodicity conditions with respect to polar coordinates would be rather awkward, if not impossible. So, planar Cartesian coordinates x,y are introduced related to the polar coordinates in the usual manner, $x=r\cos\theta, y=r\sin\theta$ and the notation $\mathbf{x} = (x, y)$ is employed: the medium is periodic with respect to $\mathbf{x}$. It is therefore useful to re-write the incremental equation on the unit cell with respect to these Cartesian coordinates when the plane-wave-expansion method is introduced in Section 3. Before this, explicit forms for the incremental moduli are derived for two popular strain energy functions for nonlinear materials.

2.3. Strain energy functions

2.3.1. The Mooney–Rivlin strain energy function

For an incompressible Mooney–Rivlin material, the strain energy function is given by

$$W = \frac{\mu}{2}\left(S_1(\lambda_r^2 + \lambda_\theta^2 + \lambda_z^2 - 3) + (1 - S_1)(\lambda_r^2\lambda_\theta^2 + \lambda_r^2\lambda_z^2 + \lambda_\theta^2\lambda_z^2 - 3)\right), \tag{8}$$

where $S_1 \in [0,1]$ is an empirically determined material constant. A neo-Hookean material corresponds to taking $S_1=1$. The Mooney–Rivlin model is typically used to describe rubber-like materials and is said to be more realistic than the neo-Hookean model at large deformations.

It has been shown that for a Mooney–Rivlin material, the shear moduli in the annular region, $\mu_r(r)$ and $\mu_\theta(r)$, are given by [39]

$$\mu_j(r) = \frac{T\mu_1}{\zeta^2}\left(1 + \frac{m}{f_j(r)}\right) \quad \text{for } r_0 < r < r_1, \tag{9}$$

with j=r,θ where $f_r(r)=r^2, f_\theta(r)=-(r^2+M)$ and the parameters m and T are defined as $T=1+(\zeta-1)S_1, m=M\zeta S_1/T$.

In the case of a neo-Hookean solid ($S_1=1$) with specific material properties, a perfect (in principal) finite cloak can be theoretically observed using the deformation described in Section 2.1, provided $\mu_1=\zeta\mu_0$ and $\rho_1=\zeta\rho_0$ [37]. This implies that the scattering behaviour in the deformed state is identical to the stress-free case. The notion is employed to give rise to two effects in Sections 4.1 and 4.2 associated with the invariance of the phononic response to pre-stress.

2.3.2. The Fung strain energy function

Since the Mooney–Rivlin strain energy function is essentially an extension of the Neo-Hookean, an alternative strain energy function is studied in order to assess its influence on the tunability of band gaps. The chosen form is a simplified version of the Fung strain energy function given by [40]

$$W = \frac{\mu_1}{2}\left(e^Q - 1\right), \tag{10}$$

where $Q = \frac{1}{2}\left(\lambda_r^2 + \lambda_\theta^2 + \lambda_z^2 - 3\right)$. This strain energy function is often used to model soft biological tissues.

The incremental shear moduli in the annular region for this case are

$$\mu_j(r) = \frac{\mu_1 \alpha_j (r^2 + M)}{2r^2}\left(\frac{\alpha_j(r^2 + M)}{r^2} + \zeta - 1\right)e^Q, \tag{11}$$

with j=r,θ, where $\alpha_r=1/\zeta, \alpha_\theta=\zeta$. As shall be seen, significant differences in incremental moduli arise due to the choice of strain energy function, a detail that manifests itself in a strongly modified material response to incremental waves.

3. PLANE-WAVE-EXPANSION (PWE) METHOD

The plane-wave-expansion method has been found to be a highly efficient technique to derive the band gap structure of periodic media [5]. In general however this method is applied to materials where each phase is homogeneous and therefore the coefficients appearing in the governing equations are usually simply piecewise-constant. Here the method is adapted to the case where material properties are permitted to be inhomogeneous *within phases* . Furthermore, it is noted that although polar coordinates are a natural representation for the geometry of the cylindrical annuli and imposition of boundary conditions, the PWE naturally accounts for any jumps in material properties and therefore what is important is to impose periodicity, which in this case is in the x andy directions. As such transformation of the incremental equation to Cartesian coordinates is necessary.

The PWE scheme represents the displacement as a sum of plane waves modulated by a Bloch phase term in order to ensure quasi-periodicity of the propagating wave. In doing this, a matrix eigenvalue problem is obtained for the permitted frequencies of a given Bloch wavenumber. In the present case, the displacement $w(\mathbf{x})$ in (6) is written as

$$w(\mathbf{x}) = e^{i\mathbf{K}\cdot\mathbf{x}} \sum_{\mathbf{G}} W[\mathbf{G}] e^{i\mathbf{G}\cdot\mathbf{x}}, \tag{12}$$

Where $\mathbf{K}$ is the Bloch wavevector, $\mathbf{G}$ are the reciprocal lattice vectors and $W[\mathbf{G}]$ are the Fourier coefficients of the displacement. The sum in (12) is over all possible reciprocal lattice vectors, which for a square lattice of period ℓ, are given by

$$\mathbf{G} = \frac{2\pi}{\ell}\left(m\mathbf{e}_x + n\mathbf{e}_y\right). \tag{13}$$

where $\mathbf{e}_x$ and $\mathbf{e}_y$ are unit vectors in the x and y directions and m and n span the integers. The sum in Eq. (12) is a double sum over both m and n.

Since periodicity is with respect to Cartesian coordinates, in order to proceed with the implementation of the PWE, Eq. (6) must be transformed. Note that (6) can be written in the form

$$\nabla \cdot (\mathbf{e}_r\sigma_{zr} + \mathbf{e}_\theta\sigma_{z\theta}) + \rho\omega^2 w = 0 \tag{14}$$

where $\mathbf{e}_r$ and $\mathbf{e}_\theta$ are unit vectors in the radial and azimuthal directions, $\sigma_{zr}=\mu_r\partial w/\partial r$and $\sigma_{z\theta}=\mu_\theta/r\partial w/\partial\theta$. It is straightforward to show that this becomes, in Cartesian coordinates,

$$\nabla \cdot (\mathbf{e}_x\sigma_{zx} + \mathbf{e}_y\sigma_{zy}) + \rho\omega^2 w = 0 \tag{15}$$

where $\sigma_{zx}=C_{zxzx}\partial w/\partial x+C_{zxzy}\partial w/\partial y,\sigma_{zy}=C_{zyzx}\partial w/\partial x+C_{zyzy}\partial w/\partial y$ and

$$C_{zxzx} = \frac{1}{x^2+y^2}(x^2\mu_r + y^2\mu_\theta). \tag{16}$$

$$C_{zxzy} = C_{zyzx} = \frac{xy}{x^2+y^2}(\mu_r - \mu_\theta). \tag{17}$$

$$C_{zyzy} = \frac{1}{x^2+y^2}(y^2\mu_r + x^2\mu_\theta). \tag{18}$$

These can now be written in the form of Fourier series:

$$C_{zxzx}(\mathbf{x}) = \sum_{\mathbf{G}} \mu_{xx}[\mathbf{G}]e^{i\mathbf{G}\cdot\mathbf{x}}, \qquad \mu_{xx}[\mathbf{G}] = \frac{1}{A}\int_{Cell} C_{zxzx}(\mathbf{x})e^{-i\mathbf{G}\cdot\mathbf{x}}d\mathbf{x}, \tag{19}$$

$$C_{zxzy}(\mathbf{x}) = \sum_{\mathbf{G}} \mu_{xy}[\mathbf{G}]e^{i\mathbf{G}\cdot\mathbf{x}}, \qquad \mu_{xy}[\mathbf{G}] = \frac{1}{A}\int_{Cell} C_{zxzy}(\mathbf{x})e^{-i\mathbf{G}\cdot\mathbf{x}}d\mathbf{x}, \tag{20}$$

$$C_{zyzy}(\mathbf{x}) = \sum_{\mathbf{G}} \mu_{yy}[\mathbf{G}] e^{i\mathbf{G}\cdot\mathbf{x}}, \qquad \mu_{yy}[\mathbf{G}] = \frac{1}{A}\int_{Cell} C_{zyzy}(\mathbf{x}) e^{-i\mathbf{G}\cdot\mathbf{x}} d\mathbf{x} \tag{21}$$

where A=ℓ^2 is the area of the periodic cell. Employing these in (15) it is easily shown that

$$\sum_{\mathbf{G}'}\Big[\sum_{\mathbf{G}} \mu_{xx}[\mathbf{G}'-\mathbf{G}](K_x+G'_x)(K_x+G_x) + \mu_{xy}[\mathbf{G}'-\mathbf{G}]((K_x+G'_x)(K_y+G_y)+(K_y+G'_y)(K_x+G_x))$$
$$+\mu_{yy}[\mathbf{G}'-\mathbf{G}](K_y+G'_y)(K_y+G_y) - \omega^2\mathcal{R}[\mathbf{G}'-\mathbf{G}]\Big]W[\mathbf{G}]e^{i(\mathbf{K}+\mathbf{G}')\cdot\mathbf{x}} = 0 \tag{22}$$

noting that $\mathbf{K} = K_x\mathbf{e}_x + K_y\mathbf{e}_y$, $\mathbf{G} = G_x\mathbf{e}_x + G_y\mathbf{e}_y$ and ρ has also been written in Fourier series form with Fourier coefficients $\mathcal{R}[\mathbf{G}]$. The exponential term is separated from the sum over $\mathbf{G}$. Since this equation must be true for any $\mathbf{x}$, it must be the case that for every $\mathbf{G}'$ the sum over $\mathbf{G}$ must be zero, i.e.

$$\Big[\sum_{\mathbf{G}} \mu_{xx}[\mathbf{G}'-\mathbf{G}](K_x+G'_x)(K_x+G_x) + \mu_{xy}[\mathbf{G}'-\mathbf{G}]((K_x+G'_x)(K_y+G_y)+(K_y+G'_y)(K_x+G_x))$$
$$+\mu_{yy}[\mathbf{G}'-\mathbf{G}](K_y+G'_y)(K_y+G_y) - \omega^2\mathcal{R}[\mathbf{G}'-\mathbf{G}]\Big]W[\mathbf{G}] = 0. \tag{23}$$

This is now seen to be in the form of a generalised eigenvalue problem and will give the full dispersion relation associated with the medium. It is conventional to scan around the outside of the irreducible Brillouin zone and pick out the lowest frequency modes [6] and [41].

4. RESULTS

When presenting results associated with dispersion relations, the band diagrams will be plotted in the usual manner. That is, the frequency of each mode is determined whilst scanning the wavenumber around the edge of the irreducible Brillouin zone, as described in [2]. Following convention, the three points defining this boundary are denoted Γ ($\hat{\mathbf{K}} = 0\mathbf{i} + 0\mathbf{j}$), X ($\hat{\mathbf{K}} = 0.5\mathbf{i} + 0\mathbf{j}$) and M ($\hat{\mathbf{K}} = 0.5\mathbf{i} + 0.5\mathbf{j}$).

Throughout the results section, the shear modulus and density are non-dimensionalised on those of the host material and the wavenumbers on the lattice period ℓ, i.e.

$$\hat{\mu}(\mathbf{x}) = \frac{\mu(\mathbf{x})}{\mu_0}, \qquad \hat{\rho}(\mathbf{x}) = \frac{\rho(\mathbf{x})}{\rho_0}, \qquad \hat{\mathbf{K}} = \ell\mathbf{K}, \qquad \hat{\mathbf{G}} = \ell\mathbf{G}. \tag{24}$$

The nondimensional frequency therefore becomes

$$\hat{\omega} = \frac{\ell\omega}{c_0}, \tag{25}$$

where $c_0 = \sqrt{\mu_0/\rho_0}$ is the shear wave speed in the host material (and unstressed annulus due to the choice of properties above). Through non-dimensionalising the spatial variable on the period, ℓ,

$$\hat{x} = \frac{x}{\ell}, \qquad \hat{y} = \frac{y}{\ell} \tag{26}$$

the unit cell is mapped to the region $-1/2 < \hat{x} < 1/2$ and $-1/2 < \hat{y} < 1/2$. The outer radius of the pre-stressed region remains fixed at $R_1 = r_1 = 0.45$. This value was chosen since it is large enough to give a sizeable pre-stressed region, but not too large so the pre-stressed region is close to the outer edge of the unit cell. Although it is possible to study the latter case as r_1 gets closer to 0.5, a significant increase in the number of terms in the PWE is required. Since this is not the main focus of the study it is avoided here.

For all calculations, the maximum plane wave number is taken at 10 (corresponding to 441 plane waves) which has been established to give sufficiently accurate results. The percentage error in the method is estimated by taking a large number of plane waves and is typically below 0.05%.

The determination of analytical results and equivalence properties associated with the neo-Hookean medium in particular allows two interesting observations to be made before the notion of configurable band gaps are considered.

4.1. Equivalent phononic response

It has already been noted that if the annular regions are chosen to be neo-Hookean with material properties related to the host via $\mu_1=\zeta\mu_0$ and $\rho_1=\zeta\rho_0$, the scattering properties of the deformed medium with deformed inner annular radii r_0 are equivalent to those of an undeformed medium with *uniform* properties μ_0 and ρ_0 and inner annular radii R_0; this property has been exploited in the study of elastodynamic cloaking for SH waves by choosing R_0 to be small and $r_0 \gg R_0$ [37]. These two materials are phononically equivalent, as illustrated in Fig. 2. Thus, the material could have quite distinct geometrical characteristics but its phononic behaviour would be equivalent. In particular, it is noted that this effect would permit a wide range of periodic materials, each medium having different cavity radii but identical phononic response.

This effect is exemplified in Fig. 3. The equivalence is illustrated in 3(a), taking a medium with $R_0=0.1$, the dispersion curves associated with the unstressed scenario where the host medium is uniform (having properties μ_0 and ρ_0) is plotted as solid lines and the pre-stressed neo-Hookean case (with annular properties $\mu_1=\zeta\mu_0$ and $\rho_1=\zeta\rho_0$) is indicated with circles. The Mooney–Rivlin case with $S_1=0.6$ is also shown to illustrate the effect that the strain energy function has on the invariance of the band structure. This reveals that we must be careful about the choice of constitutive material. In Fig. 3(b) a material with a band gap present (with R_0=0.35) is compared to an invariant material where no band gap is present, even though the deformed inner radius is also r_0=0.35.

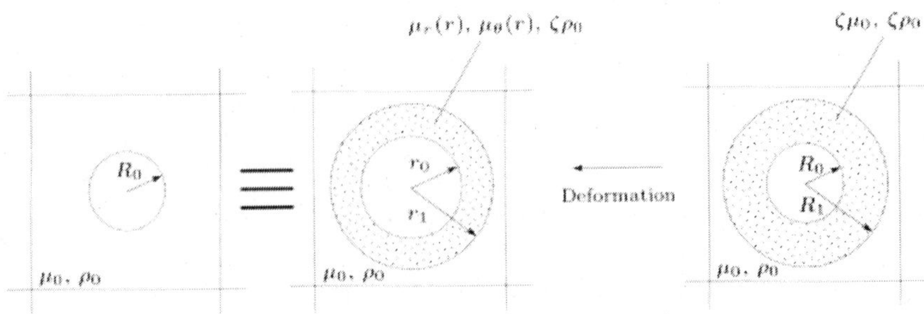

Figure. 2. Illustration of phononic equivalence. The medium on the left, consisting of a periodic array of cavities in a homogeneous medium with shear modulus μ_0 and density ρ_0 has a specific band gap structure. The medium in the middle, consisting of a periodic array of cylindrical annuli with shear modulus μ_1 and density ρ_1 embedded in a homogeneous medium with shear modulus μ_0 and density ρ_0, is phononically equivalent to the medium on the left when the annuli are pre-stressed with axial stretch ζ*if* $\mu_1=\zeta\mu_0, \rho_1=\zeta\rho_0$. The equivalent material in its undeformed state is shown on the far right.

4.2. A phononic cloak

Suppose that for some application it is required to have a medium with an array of regularly spaced cylindrical cavities of a specific radius large enough to cause band gaps. However one may wish to have such an inhomogeneous medium but *not* generate band gaps. A mechanism to achieve this is to use the invariance principle above. By choosing annular cylinders of small initial radii, we can make a material which geometrically, one would expect to experience band gaps, but due to the inhomogeneous properties it does not. If one were to choose initial radii that are small enough, the pre-stressed material with *large* radii would look almost identical to a homogeneous medium. One could term this material a *phononic cloak*.

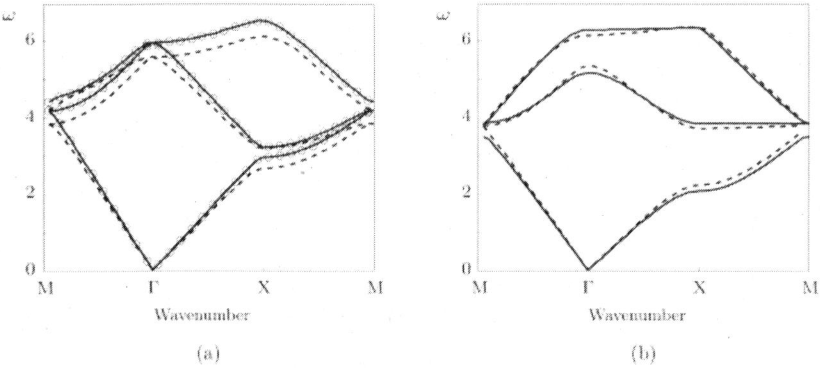

Figure. 3. (a) Dispersion curves associated with a stress-free medium (solid lines) with a square periodic array of cylindrical cavities of radius R_0=0.1. Overlaid circles are results associated with a square array of pre-stressed neo-Hookean cylindrical annuli, with R_0=0.1,ζ=1.4 and r_0=0.255 with shear modulus$\mu_1=\zeta\mu_0$ and density $\rho_1=\zeta\rho_0$. The dashed line corresponds to the response in the same pre-stressed state but when the annuli are of Mooney–Rivlin type with S_1=0.6. (b) Dispersion curves associated with a stress-free medium (solid lines) with a square periodic array of cylindrical cavities of radius R_0=0.35with a band gap evident. These are compared with dispersion curves associated with a square array of pre-stressed neo-Hookean cylindrical annuli, with R_0=0.3,ζ=1.4 and r_0=0.35 (i.e. pre-stressed to the same geometry) with shear modulus $\mu_1=\zeta\mu_0$ and density $\rho_1=\zeta\rho_0$ (dashed lines), noting that no band gap is present.

In Fig. 4 we see three different mode shapes in the cell for a stress-free (top row) and pre-stressed (bottom row) material. In the pre-stressed case we have chosen a "perfect" cloak, with R_0=0. such that even in the deformed configuration the properties are identical to a homogeneous material. It is particularly evident in Fig. 4(c) that the wave interacts significantly with the cavity in the stress-free case, but refracts around the cavity in the pre-stressed material.

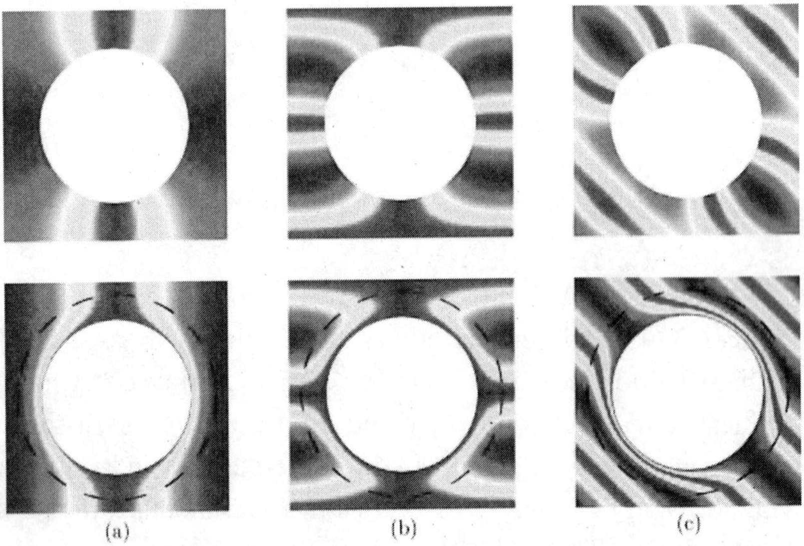

Figure. 4. Mode shapes for a periodic cell of a phononic cloak. The top row shows a stress-free homogeneous host material with cavities of radius R_0=0.33. The bottom row shows a pre-stressed material with$R_0=0,\zeta=2.2,r_0=0.33$. Material properties in the annular region shown with the dashed line are chosen to be $\mu=\zeta\mu_0$ and $\rho=\zeta\rho_0$. In (a) and (b) $k_x=\pi$, $k_y=0$, in (c) $k_x=\pi/2$, $k_y=\pi/2$.

4.3. Effective behaviour and configurable band gaps

As demonstrated, the plane-wave-expansion method determines the wave structure of the medium with any choice of linear elastic shear moduli and densities of the annulus and host phases.

However, henceforth for simplicity the properties of the host and annulus are set equal, i.e. $\mu_1=\mu_0$ and $\rho_1=\rho_0$. Before the effects of pre-stress on band gaps are considered in any detail an important aspect is studied—the influence of the pre-stress on the effective homogenised response at low frequency.

4.3.1. Effective behaviour at low frequency

The effective homogenised behaviour is obtained by determining the slope of the dispersion curves near the origin. This gives the phase velocity of the wave c_* and since it is known that at low frequency the effective density is given by $\rho_*=(1-\phi)\rho_0$ (where ϕ is the volume fraction of cavities in the deformed configuration) the effective shear modulus is thus given by $\mu_* = \rho_* c_*^2$. Of specific interest here is how the pre-stress, and consequent localised deformation within the annulus region, modifies this effective response.

We can see in Fig. 5(a) how the wavespeed varies with inner radius. The solid line shows the stress-free case where we observe that the increase of the cavity radius causes the effective wavespeed to decrease. For the neo-Hookean (dashed) and Fung (dot-dash) models there is an increase in the effective wavespeed, whereas for the Mooney–Rivlin (dotted) model, with the parameter chosen to be $S_1=0.6$, there is almost no change in the effective wavespeed. The Fung strain energy function causes a significantly larger increase in the effective wavespeed over the entire range of radii, presumably due to the increased stiffness resulting from the exponential form of the strain energy function. This shows that the choice of strain energy function has a significant effect on the low frequency homogenised behaviour of the periodic solid. Note that neither neo-Hookean or Fung has an I_2 strain invariant dependence, which is therefore assumed to generate the rather different response of the Mooney–Rivlin medium [42].

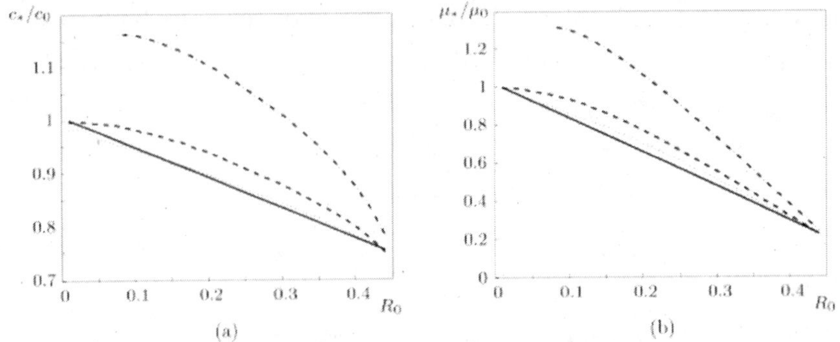

Figure. 5. The variation in (a) the effective wavespeed and (b) the effective shear modulus with undeformed cavity radius, scaled on the speed and shear modulus in the stress-free host, respectively. Solid lines show the stress-free case, dashed lines the neo-Hookean case, dotted lines the Mooney–Rivlin case (S_1=0.6) and the dot-dash lines the Fung case. In the pre-stressed materials an axial stretch of ζ=1.2 is applied.

4.3.2. Band gap tuning

We now investigate the effect of the pre-stress on the shape of the band gap with the idea of selectively activating or de-activating band gaps. Although it is possible to find interesting behaviour within the Brillouin zone [41] this behaviour is not of interest to us here since only stop bands are being sought. The band diagrams of the first three modes in a number of cases are plotted in Fig. 6 and Fig. 8. Fig. 7 shows the band gap widths for the alternative nonlinear materials considered here. These different materials are considered in turn and are then compared.

Neo-Hookean material

Consider first the simplest strain energy function—the neo-Hookean material. Shown inFig. 6(a) and (b) are the first three modes for the case when the annulus has an initial inner radius of

R_0=0.3 in the stress-free configuration and the pre-stressed configuration corresponding to ζ=1.5 (for which the deformed inner radius is r_0=0.357) respectively. It is clearly seen by comparing (a) and (b) that a stop band has been switched on by this inflation. Additionally, it is noted that the dispersion curves associated with the wavenumber regions between Γ and X for the third mode have become flatter, indicating that the pre-stress has induced regimes of very slow waves.

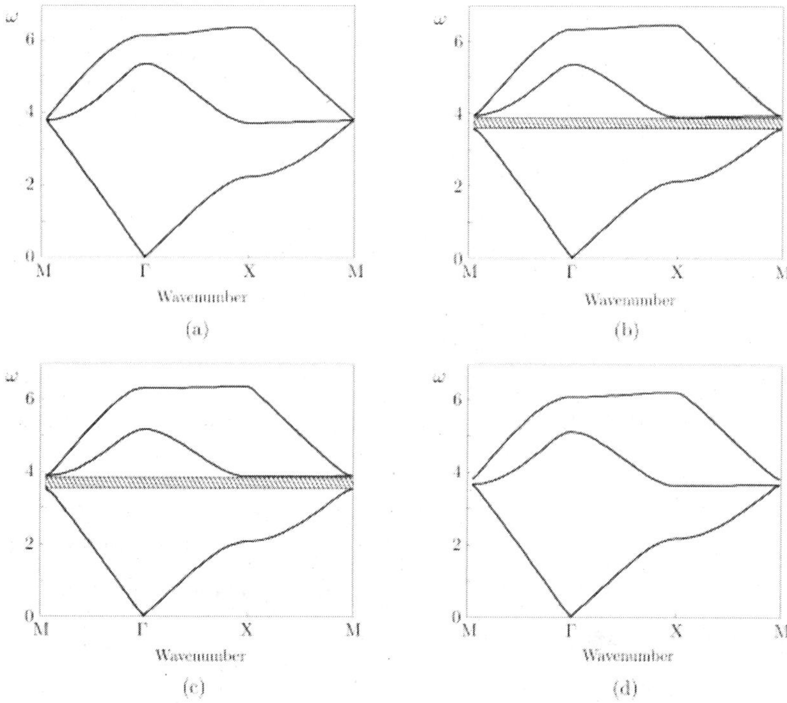

Figure. 6. The first three modes for a neo-Hookean cylinder. Stop bands are indicated by the shaded regions. Figure (a) shows the stress-free case with inner radius R_0=0.3; (b) the deformed configuration corresponding to initial inner radius R_0=0.3, an axial stretch ζ=1.5 and deformed inner radius r_0=0.357; a stop band has been "switched on". (c) The stress-free case with inner radius R_0=0.35; and (d) the deformed configuration corresponding to an initial inner radius R_0=0.35, an axial stretch ζ=0.6 and deformed inner radius r_0=0.263. The stop band has been "switched off" in this case.

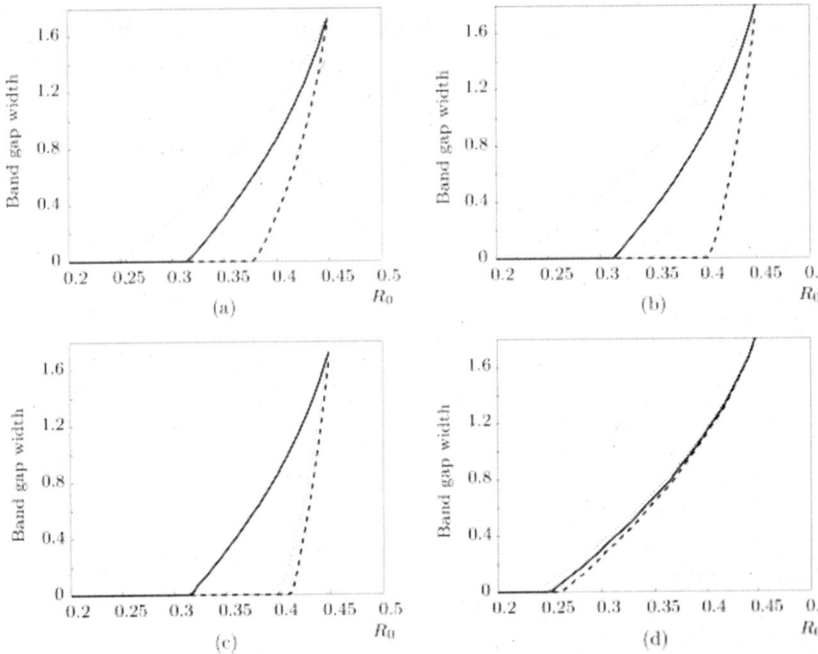

Figure. 7. The band gap width against undeformed inner radius, for various values of pre-stress in the cases of a (a) neo-Hookean medium, (b) the Mooney–Rivlin material with S_1=0.6, and (c) the Fung medium. Shown are axial stretches of ζ=1 (solid), ζ=0.8 (dash) and ζ=1.2 (dotted). In (d) we plot the band gap width of the Mooney–Rivlin strain energy function for a pre-stress of ζ=1.4 and various values of S_1 corresponding toS_1=1 (dashed), S_1=0.8 (solid) and S_1=0.6 (dotted).

In Fig. 6(c) and (d) the opposite effect is illustrated. A naturally occurring stop band in the*unstressed* material is switched off by lowering the internal pressure (or equivalently, applying an axial contraction) and decreasing the inner radius. In this case, the first three bands are plotted for an initial inner radius of the annulus of R_0=0.35 in the stress-free configuration alongside the first three bands for the pre-stressed configuration corresponding to ζ=0.6 (for which the deformed inner radius is r_0=0.263).

Fig. 7(a) shows how the width of the band gap varies with the undeformed inner radius. Illustrated are three different axial stretches, ζ=0.8 (deflation), ζ=1.2 (inflation) alongside the stress-free case ζ=1. In the stress-free case, a band gap opens up slightly above R_0=0.3 and then grows as R_0 increases. The same behaviour is experienced for the deformed configurations with larger band gaps for inflated cylinders. If one is to choose an initial radius below R_0=0.31, one could introduce a stop band by applying an appropriate amount of pre-stress.

Mooney–Rivlin material

Let us turn now to the effect of departure from neo-Hookean behaviour, by considering a Mooney–Rivlin material, reducing the parameter S_1 from unity, i.e. the neo-Hookean case. Fig. 8 illustrates the band structure resulting from an undeformed radius ofR_0=0.27 alongside that of an inflated cylinder, with an axial stretch of ζ=1.3. InFig. 8(b) the curves are plotted associated with the neo-Hookean case S_1=1, whereas in Fig. 8(c) we have a Mooney–Rivlin medium with S_1=0.5. It is seen that even with this rather large change to the constitutive behaviour (S_1 is normally closer to unity in practice) it causes very little change in the shape of the band structure when compared to the neo-Hookean case. However, it is noted that for the same deformation a stop band is switched on in the Mooney–Rivlin case which is not present for the neo-Hookean case.

The band gap widths for the Mooney–Rivlin material are plotted in Fig. 7(b) for comparison with the neo-Hookean case. In agreement with the observation from the band diagrams, the choice of a Mooney–Rivlin material increases the widths of the band gaps. Aside from that, the widths follow very similar patterns. The presence of the strain invariant I_2 can be seen to widen band gaps in general. This is illustrated in Fig. 7(d), where the band gap width is depicted for a Mooney–Rivlin material for an axial stretch ofζ=1.4 and various values of S_1. The neo-Hookean case (S_1=1) is included for comparison. It is clear that decreasing the parameter S_1 increases the band gap width for all values of the undeformed radius. The difference is most noticeable for small radii. For initial radii above R_0=0.3 there is almost no difference between having a neo-Hookean or Mooney–Rivlin material.

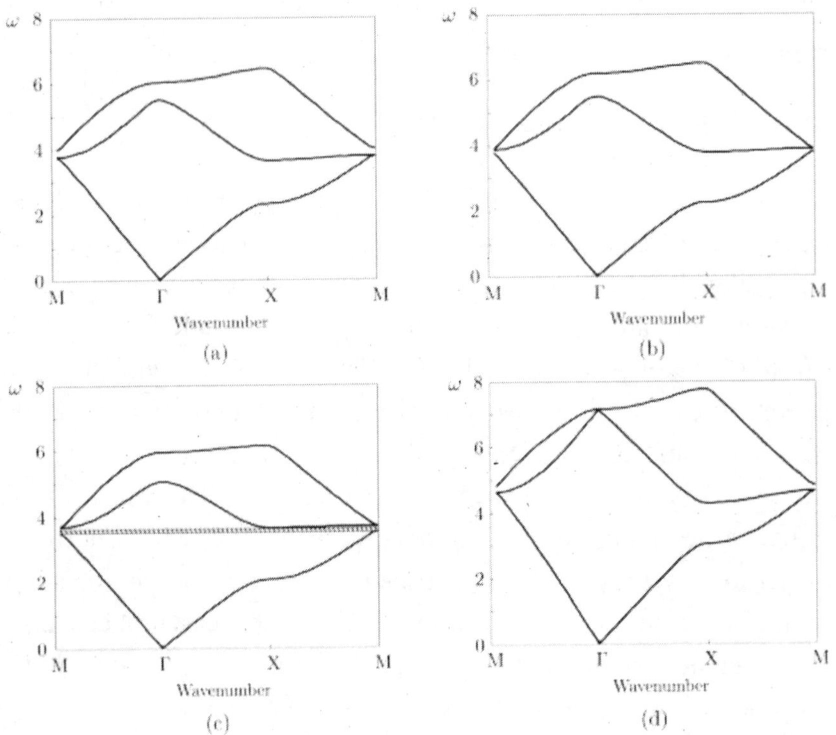

Figure. 8. The first three modes for a cylinder of initial radius R_0=0.27. Pictured are the band diagrams for (a) the stress-free case, alongside the deformed configuration with axial stretch ζ=1.3 and deformed inner radius r_0=0.32 for (b) a neo-Hookean material, (c) a Mooney–Rivlin material (S_1=0.5) and (d) a Fung-type material. Stop bands are shown by the shaded regions.

Fung type material

In Fig. 8(d) the equivalent band diagram for a Fung material is plotted alongside the neo-Hookean and Mooney–Rivlin materials for an initial radius of $R_0=0.27$ and axial stretch$\zeta=1.3$. Clearly using a Fung material has a significant effect on the lowest three modes, raising them to a higher frequency. This is likely due to the Fung material stiffening up significantly when large deformations are applied.

It is interesting to note that the stop band is not present for the Fung material, due to the significant change in the band gap shape. In particular, the frequency of the second band at point X is lower (in relation to the frequency at point M) when compared with the stress-free case. This appears to stop the band gap opening up.

Fig. 7(c) gives the band gap width for the Fung material as a function of undeformed radius. What is most interesting is that, contrary to the other materials, inflating a cavity in a Fung material always has the effect of decreasing the width of a stop band. This appears to happen for any amount of stretch. This highlights the point that the constitutive behaviour of the chosen material is extremely important. The Fung medium can in fact be seen as an extreme case of an 'I_1 only model'.

Influence of the choice of strain energy function

The difference between the strain energy functions considered here can be understood by analysing the inhomogeneous shear moduli. An increase in the S_1 towards unity (i.e. approaching the neo-Hookean case) accentuates the effect that the pre-stress has on the shear moduli. The neo-Hookean case is clearly the limit (in the context of the class of Mooney–Rivlin materials) at which there is maximum effect on the shear moduli, explaining why the effective wavespeed is so much higher in the neo-Hookean material. Due to the exponential nature of the Fung strain–energy function, the pre-stress has a much more significant effect on the shear moduli in comparison with the Mooney–Rivlin case. This leads to a more pronounced change in the band diagram as seen in Fig. 8(d). This behaviour can be associated with the I_1 only dependence and the Fung model is clearly an extreme example of such a model, given its exponential form.

4.4. Further observations

Further to the opening of stop bands, other more general observations can be made about the band diagrams. It is interesting to note that in both Figs. 3(a) and 8(d) the second and third bands are touching at the wavenumber Γ. In the Fung case, these intersecting bands are separated by a significant distance before the application of the pre-stress.

A final and important observation to be made is that the effect of the pre-stress goes further than just a change in the geometry. This can be seen from Fig. 9 in which the band diagram is plotted for a stress-free material alongside a pre-stressed medium, where the inner radius of the former is equal to the deformed inner radius of the latter. Although the shape of the bands are similar in some regions, there are several significant differences. For example, the difference in frequency at the wavenumber Γ is large when comparing the pre-stressed and equivalent material. This disparity implies that the inhomogeneity in the shear moduli caused by the deformation has a significant effect on the band structure.

5. CONCLUSION

A new geometrical design has been proposed in order to tune the band gap structure of a periodic elastic composite using nonlinear elastic pre-stress, thus permitting the design of phononic switches for antiplane elastic waves. In particular, the onset radius of the acoustic (lowest frequency) stop band has been investigated. It is found that applying an inflation inside the annulus tends to cause the stop band to switch on whereas a deflation tends to switch the band gap off. It was shown that the choice of material (via a variation of the strain energy function) has a significant influence on the size of the stop band. The concepts of phononic invariance and phononic cloaking were also introduced, in principal enabling a range of designs of materials with specific geometries and tunable band structures.

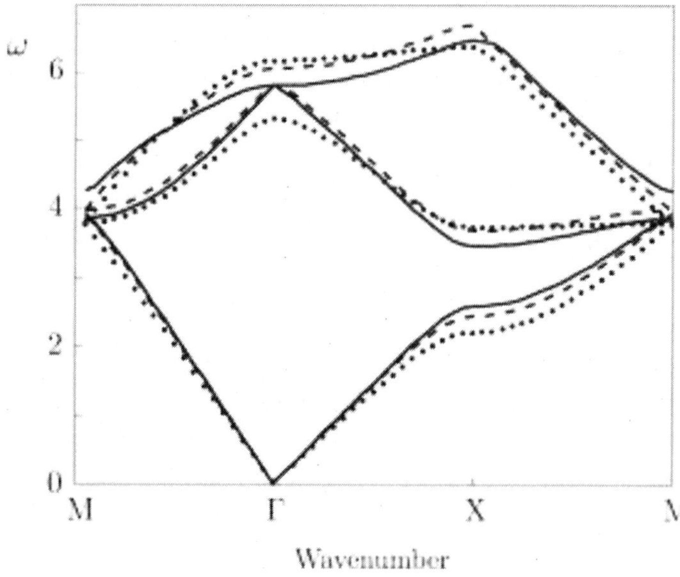

Figure. 9. The first three bands in the stress-free configuration with inner radius R_0=0.2 (solid line), the stress-free configuration with inner radius R_0=0.3 (dotted line) and the pre-stressed configuration with initial inner radius R_0=0.2, axial stretch ζ=1.5 and deformed inner radius R_0=0.3 (dashed line). This plot reveals that the pre-stress has an effect which goes further than just a change in geometry.

The approach utilised here is easily extendable to other configurations and more complex geometries, and indeed it is expected that this will provide a rich variety of band gap tuning strategies. In determining the band gap structure the plane-wave-expansion method was modified and extended in order to deal with material coefficients that are generally inhomogeneous, not simply piecewise constant as is almost always the case in standard problems. This extension will apply analogously to the coupled two- and three-dimensional elastodynamic problems, albeit with an obvious increase in complexity and system size.

Since the plane-wave-expansion method is implemented numerically, there are no restrictions on the geometries considered except for a limit on the size of the outer radius of the deformed cylinder; if the latter is too large then the cylinders get close to touching. In this case a large number of plane waves are required for convergence and although solvable, the computation becomes prohibitively slow on a standard desktop computer.

The only technical constraint of the present approach is our ability to find the incremental wave equation. Work is currently under way to investigate compressible hyperelastic materials, in which case an analytical solution for the nonlinear deformation is generally not possible; hence, a nonlinear ordinary differential equation must be solved numerically in order to determine the deformation. This will then necessitate a numerical approach to finding the incremental inhomogeneous moduli appearing in the incremental wave equation. For more complex geometries and loadings a numerical approach will also be required to determine the static deformation, associated stress fields, and incremental moduli.

ACKNOWLEDGEMENTS

The authors acknowledge the Leverhulme Trust for financial support via grantF/00120/CC which funded the PhD studentship of Barnwell. Parnell is grateful to the Engineering and Physical Sciences Research Council for funding his research fellowship (EP/L018039/1) and Abrahams thanks the Royal Society for a Wolfson Research Merit award (2013–2018).

REFERENCES

1. R. Craster, S. Guenneau, Acoustic Metamaterials: Negative Refraction, Imaging, Lensing and Cloaking, Springer, 2013.

2. P. Deymier, Acoustic Metamaterials and Phononic Crystals, Springer, 2013.

3. M. Sigalas, E.N. Economou, Elastic and acoustic wave band structure, J. Sound Vib. 158 (2) (1992) 377–382.

4. M. Sigalas, E.N. Economou, Band structure of elastic waves in two dimensional systems, Solid State Commun. 86 (30) (1993) 141–143.

5. M.S. Kushwaha, P. Halevi, L. Dobrzynski, B. Djafari-Rouhani, Acoustic band structure of periodic elastic composites, Phys. Rev. Lett. 71 (13) (1993) 2022–2025.

6. M.S. Kushwaha, P. Halevi, G. Martinez, Theory of acoustic band structure of periodic elastic composites, Phys. Rev. B: Condens. Matter Mater. Phys. 49 (4) (1994) 2313–2322.

7. M.I. Hussein, M.J. Leamy, M. Ruzzene, Dynamics of phononic materials and structures: Historial origins, recent progress and future outlook, Appl. Mech. Rev. 66 (2014) 040802.

8. C. Goffaux, J.P. Vigneron, Theoretical study of a tunable phononic band gap system, Phys. Rev. B: Condens. Matter Mater. Phys. 64 (7) (2001) 5118–5122.

9. X. Li, F. Wu, H. Hu, S. Zhong, Y. Liu, Large acoustic band gaps created by rotating square rods in two-dimensional periodic composites, J. Phys. D: Appl. Phys. 36 (1) (2003) 15–17.

10. S.C.S. Lin, T.J. Huang, Tunable phononic crystals with anisotropic inclusions, Phys. Rev. B: Condens. Matter Mater. Phys. 83 (17) (2011) 4303–4312.

11. A. Khelif, Two-dimensional phononic crystal with tunable narrow pass band: Application to a waveguide with selective frequency, J. Appl. Phys. 94 (3) (2003) 1308–1311.

12. Y. Pennec, B. Djafari-Rouhani, J.O. Vasseur, A. Khelif, P.A. Deymier, Tunable filtering and demultiplexing in phononic crystals with hollow cylinders, Phys. Rev. E (3) 69 (4) (2004) 6608–6613.

13. A. Evgrafov, C.J. Rupp, M.L. Dunn, K. Maute, Optimal synthesis of tunable elastic wave-guides, Comput. Methods Appl. Mech. Engrg. 198 (2) (2008) 292–301.

14. X. Zhang, Z.-Q. Zhang, L.-M. Li, C. Jin, D. Zhang, B. Man, B. Cheng, Enlarging a photonic band gap by using insertion, Phys. Rev. B: Condens. Matter Mater. Phys. 61 (3) (2000) 1892.

15. O.B. Matar, J.F. Robillard, J.O. Vasseur, A.C. Hladky-Hennion, P.A. Deymier, P. Pernod, V. Preobrazhensky, Band

gap tunability of magneto-elastic phononic crystal, J. Appl. Phys. 111 (5) (2012) 4901–4915.

16. J.Y. Yeh, Control analysis of the tunable phononic crystal with electrorheological material, Physica B 400 (1) (2007) 137–144.

17. W. Yang, L. Chen, The tunable acoustic band gaps of two-dimensional phononic crystals with a dielectric elastomer cylindrical actuator, Smart Mater. Struct. 17 (2008) 015011.

18. Z. Hou, F. Wu, Y. Liu, Phononic crystals containing piezoelectric material, Solid State Commun. 130 (11) (2004) 745–749.

19. X.Y. Zou, Q. Chen, B. Liang, J.C. Cheng, Control of the elastic wave bandgaps in two-dimensional piezoelectric periodic structures, Smart Mater. Struct. 17 (1) (2008) 5008–5012.

20. F. Casadei, T. Delpero, A. Bergamini, P. Ermanni, M. Ruzzene, Piezoelectric resonator arrays for tunable acoustic waveguides and metamaterials, J. Appl. Phys. 112 (6) (2012) 4902–4906.

21. X. Zhou, C. Chen, Tuning the locally resonant phononic band structures of two-dimensional periodic electroactive composites, Physica B 431 (1) (2013) 23–31.

22. Z.G. Huang, T.T. Wu, Temperature effect on bandgaps of surface and bulk acoustic waves in two-dimensional phononic crystals, IEEE Trans. Ultrason. Ferroelectr. Freq. Control 52 (3) (2005) 365–370.

23. K.L. Jim, C.W. Leung, S.T. Lau, S.H. Choy, H.L.W. Chan, Thermal tuning of phononic bandstructure in ferroelectric ceramic/epoxy phononic crystal, Appl. Phys. Lett. 94 (19) (2009) 3501–3503.

24. J.Y. Ou, E. Plum, L. Jiang, N.I. Zheludev, Reconfigurable photonic metamaterials, Nano Lett. 11 (2011) 2142–2144.

25. W. Zhu, Y. Huang, I.D. Rukhlenko, G. Wen, M. Premaratnel, Configurable metamaterial absorber with pseudo wideband spectrum, Opt. Express 20 (6) (2012) 6616–6621.

26. G.A. Rogerson, K.J. Sandiford, The effect of finite primary deformations on harmonic waves in layered elastic media, Internat. J. Solids Structures 37 (14) (2000) 2059–2087.

27. W.J. Parnell, Effective wave propagation in a pre-stressed nonlinear elastic composite bar, IMA J. Appl. Math. 72 (2) (2007) 223–244.

28. G. Shmuel, G. deBotton, Band-gaps in electrostatically controlled dielectric laminates subjected to incremental shear motions, J. Mech. Phys. Solids 60 (11) (2012) 1970–1981.

29. M. Gei, A.B. Movchan, D. Bigoni, Band-gap shift and defect induced annihilation in prestressed elastic structures, J. Appl. Phys. 105 (6) (2009) 3507–3513.

30. R. Feng, K. Liu, Tuning the band-gap of phononic crystals with an initial stress, Physica B 407 (12) (2012) 2032–2036.

31. L. Wang, K. Bertoldi, Mechanically tunable phononic band gaps in three-dimensional periodic elastomer structures, Internat. J. Solids Structures 49 (19) (2012) 2881–2885.

32. K. Bertoldi, M.C. Boyce, Wave propagation and instabilities in monolithic and periodically structured elastomeric materials undergoing large deformations, Phys. Rev. B: Condens. Matter Mater. Phys. 78 (18) (2008) 4107–4122.

33. J.H. Jang, C.Y. Koh, K. Bertoldi, M.C. Boyce, E.L. Thomas, Combining pattern instability and shape-memory hysteresis for phononic switching, Nano Lett. 9 (5) (2009) 2113–2119.

34. R.W. Ogden, Non-linear Elastic Deformations, Dover, 1998.

35. M. Destrade, G. Saccomandi, Waves in Nonlinear Pre-stressed Materials, Springer, New York, 2007.

36. W.J. Parnell, Nonlinear pre-stress for cloaking from antiplane elastic waves, Proc. R. Soc. A: Math. Phys. Eng. Sci. 468 (2138) (2012) 563–580.

37. W.J. Parnell, A. Norris, T. Shearer, Employing pre-stress to generate finite cloaks for antiplane elastic waves, Appl. Phys. Lett. 100 (17) (2012) 171907.

38. W.J. Parnell, I.D. Abrahams, Antiplane wave scattering from a cylindrical void in a pre-stressed incompressible neo-Hookean material, Commun. Comput. Phys. 11 (2) (2012) 367–382.

39. T. Shearer, Waves in nonlinear elastic media with inhomogeneous pre-stress (Ph.D. thesis), University of Manchester, 2012.

40. Y.C. Fung, Biomechanics: Mechanical Properties of Living Tissues, Springer-Verlag, 1993.

41. R. Craster, T. Antonakakis, M. Makwana, S. Guenneau, Dangers of using the edges of the Brillouin zone, Phys. Rev. B 86 (11) (2012) 5130–5135.

42. C.O. Horgan, M.G. Smayda, The importance of the second strain invariant in the constitutive modeling of elastomers and soft biomaterials, Mech. Mater. 51 (2012) 43–52.

CHAPTER 10

A model of acoustic absorption in fluids based on a continuous distribution of relaxation times

G. Vilensky, G. ter Haar, N. Saffari

ABSTRACT

This work extends the quasi-equilibrium relaxation theory of sound absorption in liquids to the case of continuous distribution of relaxation times. Such extension is needed when absorption mechanisms are not confined to the action of viscosity and heat conduction, but are mainly due to the excitation of a large number of internal molecular degrees of freedom. In this case the conventional Navier–Stokes equations are not sufficient to describe the fluid motion, and additional equations are required to model normal relaxation stresses. When relaxation frequencies form a sufficiently dense distribution, as is the case for many biological fluids, it makes sense to consider the limit of continuously distributed relaxation frequencies, in order to obtain the required equation for normal relaxation stresses.

In contrast to its discrete counterparts, the proposed method avoids the use of a potentially infinite number of relaxation equations for a given set of distinct relaxation frequencies. Instead, these are replaced by a single evolution equation of Boltzmann type whose right-hand side is a linear combination of the time derivatives of density and entropy. The rheological functions appearing before these derivatives are expressed in terms of the absorption coefficient. Since the dependence of absorption coefficient on sound frequency is measurable experimentally, these rheological coefficients can be recovered from experimental data.

The key feature of the present study is that a closed system of equations of motion can be formulated directly from absorption measurement data on the basis of the theory proposed for the very wide range of absorption laws that can occur in practice.

As an illustration of the generality of the present method, a number of absorption laws documented in the experimental literature are considered in detail, in order to derive the coefficients of the related systems of equations of motion for these liquids. For example, the methodology based on modelling of acoustic absorption in biologically soft tissue by the employment of fractional derivatives, which has been recently developed in the literature, is shown to be a special case of the proposed theory.

HIGHLIGHTS

- Sound absorption theory in fluid with continuous distribution of relaxations is given.
- The generalization of the discrete relaxation theory of sound absorption is given.
- The fractional dynamics equation for sound waves in bio-tissue is its special case.

- A method to reconstruct equations of motion from experimental data is given.

KEYWORDS

Sound absorption; High intensity focused ultrasound; Relaxation losses; Biological fluids; Soft tissue

1. INTRODUCTION

This paper is concerned with the problem of sound absorption in liquids with complex molecular structure. The main motivation for this study is the need to design high intensity focused ultrasound (HIFU) equipment for the treatment of oncological disease. The related medical procedure relies on the ability to convert the energy of an ultrasound beam, which is focused on a tumour, into a thermal field whose temperature is sufficient to denature cellular proteins. The central mechanism involved in this process is ultrasound absorption.

When dissipation due to the relaxation mechanisms associated with the redistribution of the energy of macroscopic motion into the internal degrees of freedom of molecular motion is weak, acoustic losses can be accounted for by conventional viscosities and heat conductivity. The related small amplitude linear theory based on the Navier–Stokes model then gives a quadratic dependency of sound absorption coefficient on frequency,[1], [2] and [3].

However, for example, in biological tissue the effects of viscosity and heat conduction are small compared to relaxation losses (see [4] and the references therein). Experiments have shown that, for a wide range of frequencies, dependence of absorption on frequency obeys a power law whose exponent is close to unity, rather than the conventional quadratic law [5], [6], [7] and [8].

For several decades this constituted the major difficulty in formulating the time-domain wave equation needed, for example, for efficient numerical modelling of problems with realistic geometries. A method of constructing such a wave equation was proposed by [9]. In this paper Szabo constructs a dispersion relation with the required properties in the frequency domain by the analysis of related wave operators and causality considerations. A linear wave equation in the time domain was then recovered from the dispersion relation via a Fourier inversion. A modification of Szabo's wave operator, originally obtained in terms of generalized functions, has been proposed in [10], where it was recast as an integro-differential operator where the lossy term was represented by a fractional derivative with respect to the time variable. A generalization of this model, for the case in which attenuation results from the spatial microstructure of the propagation medium, is presented in [11]. In contrast to the time-varying Caputo fractional derivative used originally by them in Ref. [10], in [11] Chen and Holm use a fractional Laplacian, which involves a space-varying fractional derivative. Further development of this approach and its numerical implementation can be found in [12].

A fractal viscoelastic model with an infinite number of alternating nested springs and dash-pots was recently proposed [13] in order to give physical justification for the power law absorption

equation (1)

$\alpha_{ab} \sim |\omega|^{\chi}$,

where ω is the sound frequency. In this approach permissible values of the power law exponent χ were found to be given by the expression

equation (2)

$$\chi = \frac{3}{2} - \frac{1}{2N+1} + \frac{1}{2M+1},$$

where integer numbers N and M define the depth of recursion of the dampers and the springs, respectively. In the linear limit the fractional wave equation obtained in [12]was recovered. The major drawback of this model, however, is that it does not allow for the power law exponent χ to take arbitrary real values such as, for example, $\chi = \sqrt{3}, \pi/2$, *etc* . Instead, the experimentally measured power law exponent χ must be represented by the nearest value allowed by the right-hand side of the above formula. Consequently, the refinement of the experimental value of χ is not parallelled by the refinement of the governing equation. The choice of numbers N and M may also be non-unique, *i.e.* , different configurations of dash-pots and springs may result in the same law of absorption. For example, χ=1.5 for any N=M.

These difficulties can be overcome if fractional time derivatives in the relationship between the stress and strain tensors are used (see [14] and the references therein). This makes it possible to account for an arbitrary power law exponent.

However, if the absorption coefficient of a liquid does not obey the power law, as is often the case, say, in the food industry, [15], then a more generic approach may be required. The large variety of absorption laws for different applications in a number of liquids shows the need for a unifying approach to the problem.

The aim of the present work is to formulate a phenomenological theory of sound absorption in liquids from the standpoint of relaxation theory, with a continuous distribution of relaxation frequencies.

The relaxation theory of absorption and dispersion of acoustic waves was proposed for the first time by [16] (see also [17]). There appears to be no comprehensive recent review of the different relaxation theories applied to acoustics, but a detailed treatment of the subject can be found in [18].

The derivation of a linear wave equation for an inhomogeneous medium with absorption dominated by a multiple, but finite, number of relaxation processes was proposed in [4]. Its analysis builds on the extensive previous research in the area summarized in [18] and [1] and more recently [3]. The authors of [4] employ the irreversible thermodynamic theory used to describe chemical relaxations in an electrolyte.

Experiments carried out in tissues such as liver, and in protein solutions, demonstrate that the major part of the ultrasound absorption arises at a molecular, rather than cellular, or higher, level of organization, [6] and [5]. There exists a broadband statistical distribution of relaxation frequencies over the four orders of magnitude starting at 0.1 MHz, i.e. over the whole range of frequencies available for observations. Consequently, it is argued in the literature, [19], [6], [5] and [3] that continuously distributed relaxations should be used to model relaxation losses, since the assumption of a finite number of relaxation processes is not sufficient to derive a power law of absorption.

In a wider scientific context, the assumption of an infinite number of relaxation processes has been used to derive the power law relationship for a number of physical problems other than acoustics. The details of such models and of those key aspects not discussed here can be found in [20], [21], [22] and [23].

In this paper we construct such a physical model with a continuous distribution of relaxations. It is based on the phenomenological theory of relaxation processes which have a long history in physics literature and was recently summarized in a monograph in which references to other relevant publications can be found, [24]; also see [25]. The present work is confined to relaxation mechanisms which result from changes in normal stresses. More specifically, we are interested in the local mechanisms of irreversible energy loss caused by uniform compression or expansion of a medium for which all components remain unchanged, rather than the losses caused by friction between different layers of a medium which move with different velocities (for a more detailed discussion of this issue see [26]). No attempt is made to model effects of shear viscosity and heat conduction beyond the conventional Navier–Stokes approach, since this topic goes far beyond the scope of this paper.

Section 2 summarizes the related thermodynamic and relaxation theories used in Section 3 to derive the system of equations with continuously distributed relaxations. Section 4 derives the expressions for the unknown phenomenological coefficients of these systems in terms of the frequency-domain absorption coefficient which may be determined by experiment. This is the main result of the work, since it makes it possible to formulate a closed system of equations of motion in physical time-space domain directly from experimental data for absorption measurements. Section 5 provides examples of application of the proposed theory. In particular, it considers the model of absorption in soft tissue with continuously distributed relaxations. The above mentioned fractional derivative model originally discovered from analysis of dispersion relations, [9], is recovered as a special case of the approach proposed in the present work which, we believe, provides a more fundamental treatment of this problem.

2. RELAXATION EQUATIONS

The basis for the different relaxation theories of absorption is often hidden in the underlying thermodynamic theory assumptions. In the absence of a well developed statistical theory for liquids with complex molecular structure, a number of phenomenological non-equilibrium thermodynamic models has been proposed to date. In this paper we examine the approach which builds upon the original quasi-equilibrium thermodynamic model used by [16] and [17], in order to derive a closed system of governing equations.

This model assumes that every sufficiently small but macroscopic fluid element is locally in a state of thermodynamic equilibrium. The Gibbs equation is postulated as the definition of entropy s so that for a homogeneous medium

equation (3)

$$T\,ds = de + p\,dv,$$

where v=1/ρ and e are the specific volume and the internal energy per unit mass, ρis the density. The pressure, p, and the temperature, T, retain their conventional meanings.

The parameters of the relaxation processes do not enter the entropy Eq. (3), or other thermodynamic relations explicitly. In a quasi-equilibrium approach relaxations are taken into account via an equation of state.

Apart from the usual thermodynamic parameters such as ρ and s each elementary macroscopic volume of a non-equilibrium medium is characterized by a set of internal parameters $\xi_k(t)$, where k is a positive number. Each parameter $\xi_k(t)$ describes the related k-th internal molecular relaxation process. During this process the parameter $\xi_k$recovers its equilibrium value $\xi_k^o = \xi_k^o(\rho, s)$. The difference $\xi_k(t) - \xi_k^o$ characterizes the departure of the medium from equilibrium and is assumed to be small. The time derivative of the parameter $\xi_k(t)$ is then expanded in a power series in $(\xi_k - \xi_k^o)$. Since at equilibrium $d\xi_k/dt$ must be zero and $\xi_k(t)$ must remain finite at all times, this results in the relaxation equation (for details see [17])

equation (4)

$$\frac{d\xi_k}{dt} = -\frac{1}{\tau_k}(\xi_k - \xi_k^o) + O((\xi_k - \xi_k^o)^2),$$

where τ_k is the relaxation time. In general, τ_k depends on the temperature T and pressure p. Here we assume that the relaxation parameters ξ_k can be chosen so that linear interaction terms of the form

equation (5)

$$\sum_{j \neq k} \frac{1}{\tau_{kj}}(\xi_j - \xi_j^o)$$

which, in general, can appear on the right-hand side of the Taylor series expansion (4)are removed via the diagonalization procedure, so that only the k-th term of this sum is present. Hence, if direct interaction between any two relaxation processes j and ktakes place, it must be a higher order effect.

The equations governing the behaviour of the normal stresses in the presence of relaxation processes can be obtained using the methodology developed in [16], [17],[2] and [27].

For the k-th relaxation process, we define the pressure increment $p_k^r(\rho, s, \xi_k)$ over the equilibrium pressure p(ρ,s). For small perturbations of the relaxation parameter $\xi_k$from its equilibrium value ξ_k^o the pressure increment $p_k^r(\rho, s, \xi_k)$ can be expanded into the power series in $(\xi_k - \xi_k^o)$, and with the same accuracy as for the relaxation equation (4) we can write

equation (6)

$$p_k^r(\rho, s, \xi_k) = a_k(\rho, s)(\xi_k - \xi_k^o) + O((\xi_k - \xi_k^o)^2),$$

where $a_k(\rho,s)$ is an unknown coefficient. The zeroth order term is absent in (6), since $p_k^r = 0$ at equilibrium.

Since the theory is linear in $(\xi_k - \xi_k^o)$, the total pressure p_Σ is equal to the quasi-equilibrium pressure $p(\rho,s)$ plus the sum of pressure increments due to each relaxation process

equation (7)

$$p_\Sigma = p(\rho, s) + \sum_{(k)} p_k^r(\rho, s, \xi_k).$$

We also introduce the relaxation pressure $p_k(\rho,s,\xi_k)$ of the k-th relaxation process, that is, the pressure which would exist in the absence of all other relaxation processes:

equation (8)

$$p_k(\rho, s, \xi) = p(\rho, s) + p_k^r(\rho, s, \xi).$$

Taking the time derivative of (6) we obtain the equation

equation (9)

$$\frac{d}{dt} p_k^r = m_\rho^k \frac{d\rho}{dt} + m_s^k \frac{ds}{dt} + a_k \frac{d\xi_k}{dt}.$$

Since $a_k(\xi_k - \xi_k^o) = p_k(\rho, s, \xi) - p(\rho, s)$, here

equation (10)

$$m_\rho^k \equiv \frac{\partial}{\partial \rho}[a_k(\xi_k - \xi_k^o)]_s = \left(\frac{\partial p_k}{\partial \rho} - \frac{\partial p}{\partial \rho}\right)_s,$$

equation (11)

$$m_s^k \equiv \frac{\partial}{\partial s}[a_k(\xi_k - \xi_k^o)]_\rho = \left(\frac{\partial p_k}{\partial s} - \frac{\partial p}{\partial s}\right)_\rho .$$

The right-hand side of (10) is the difference of the square of the local speed of sound in the medium when only the k-th relaxation process is excited, and the square of the local speed of sound in the absence of relaxation phenomena, respectively. As shown in [17], according to Le Chatelier's principle, $m_\rho^k > 0$. The right-hand side of (11) is proportional to the difference between the rate of temperature change per unit volume expansion when only the k-th relaxation process is excited and the rate of temperature change per unit volume expansion in the absence of relaxation phenomena.

It follows from (4) and (6) that

equation (12)

$$a_k \frac{\mathrm{d}\xi_k}{\mathrm{d}t} = -\frac{a_k}{\tau_k}(\xi_k - \xi_k^o) = -\frac{1}{\tau_k} p_k^r .$$

Substitution of expression (12) into (9) results in the following equations for the pressure increments p_k^r

equation (13)

$$\frac{\mathrm{d}}{\mathrm{d}t} p_k^r = m_\rho^k \frac{\mathrm{d}\rho}{\mathrm{d}t} + m_s^k \frac{\mathrm{d}s}{\mathrm{d}t} - \frac{1}{\tau_k} p_k^r .$$

Coefficients m_ρ^k and m_s^k are functions of ρ,s,τ_k and ξ_k. However, since

equation (14)

$$m_\rho^k = -a_k \frac{\partial \xi_k^o}{\partial \rho} + \frac{\partial a_k}{\partial \rho}(\xi_k - \xi_k^o) \quad \text{and} \quad m_s^k = -a_k \frac{\partial \xi_k^o}{\partial s} + \frac{\partial a_k}{\partial s}(\xi_k - \xi_k^o).$$

to leading order, the dependence on ξ_k can be neglected and the coefficients m_ρ^k and m_x^k can be treated as the phenomenological functions to be determined by experiment.

The governing system can be closed if Eqs. (7) and (13) are supplemented by the equation of continuity, compressible Navier–Stokes equations, energy equation and the caloric equation of state relating the specific internal energy e, density and temperature [17] and [28]:

equation (15)

$$\frac{d\rho}{dt} + \rho \mathrm{div}(\underline{v}) = 0,$$

equation (16)

$$\rho \frac{dv_i}{dt} = -\frac{\partial}{\partial x_i} p_\Sigma + \frac{\partial}{\partial x_k} \sigma'_{ik},$$

equation (17)

$$\rho \frac{de}{dt} = -p_\Sigma \mathrm{div}(\underline{v}) + \sigma'_{ik} \frac{\partial v_i}{\partial x_k} + \mathrm{div}(\kappa \nabla T),$$

equation (18)

e=e(ρ,T),

where $\underline{v} = (v_1, v_2, v_3)$ is the velocity vector, $\underline{r} = (x_1, x_2, x_3)$ is the position vector, σ'_{ik} is the viscous stress tensor defined by Eq. (19)

equation (19)

$$\sigma'_{ik} = \eta \left(\frac{\partial v_i}{\partial x_k} + \frac{\partial v_k}{\partial x_i} - \frac{2}{3} \delta_{ik} \frac{\partial v_l}{\partial x_l} \right) + \zeta \delta_{ik} \frac{\partial v_l}{\partial x_l},$$

η is the shear viscosity, ς is the volume viscosity, κ is the coefficient of heat conductivity. The summation operation is carried out with respect to the repeating subscripts.

For complex media like biological fluids and soft tissue the structure of the caloric equation of state (18) is largely unknown to-date. For the purposes of the present work, it is convenient to use the following well-known thermodynamic relationship

equation (20)

$$de = \left(p - \frac{\rho T \beta c_s^2}{\gamma}\right) \frac{1}{\rho^2} d\rho + c_v dT,$$

where β is the coefficient of thermal expansion, c_s is the speed of sound at constant entropy, $\gamma = c_p/c_v$ is the specific heat ratio, c_v and c_p are the specific heats at constant volume and pressure, respectively.

No linearity assumptions are made about the quasi-equilibrium thermodynamic equation of state $p(\rho,s)$. It is taken in generic form here, since in biological fluids it is often the thermodynamic equation of state that is the main source of non-linearity.

3. CONTINUOUSLY DISTRIBUTED RELAXATIONS

To extend the above theory to continuous distribution of relaxation times, instead of p_k^r, m_ρ^k and m_s^k, it is convenient to introduce the functions:

equation (21)

$$\hat{p}^r(\varpi) = \varpi^{-1} p_k^r, \qquad \hat{m}_\rho(\varpi) = \varpi^{-1} m_\rho^k, \qquad \hat{m}_s(\varpi) = \varpi^{-1} m_s^k,$$

where $\varpi=1/\tau_k$ is a relaxation frequency. As opposed to the discrete set of parameters$\{\tau_k\}$, we now assume that $\tau_k>0$ is a continuous variable. Thus the variable ϖ is allowed to vary continuously also. Hence, every relaxation process is characterized by the relaxation frequency ϖ, and we shall now consider the relaxation pressure increment per unit relaxation frequency $\hat{p}^r$, instead of the relaxation pressure increment p_k^r.

Assume everywhere in what follows that the variations of the relaxation times $\tau_k(T,p)$due to temperature and pressure changes are negligible, and ϖ can be considered as a parameter independent of the macroscopic time t and the position vector. Eq. (13) can then be rewritten as

equation (22)

$$\frac{\partial \hat{p}^r}{\partial t} + \underline{v} \cdot \nabla \hat{p}^r + \varpi \hat{p}^r = \hat{m}_\rho \frac{d\rho}{dt} + \hat{m}_s \frac{ds}{dt}.$$

The summation in Eq. (7) is substituted by the integral over the entire range of relaxation frequencies ϖ:

equation (23)

$$p_\Sigma = p(\rho, s) + \int_0^\infty \hat{p}^r(\varpi, t, \underline{r}) d\varpi.$$

Note that the function p_Σ does not depend on ϖ, since ϖ enters the last formula as an integration variable. Neither do any of the unknown variables $\underline{v}, \rho, e, s$ or T in the governing system , , , , (19) and (20) depend on ϖ, since the only ϖ-dependent function $\hat{p}^r$ appears in this system via the function p_Σ given by Eq. (23), that is, in the integral with respect to ϖ. Consequently, if we define functions $p^r(\tau)$,$m_\rho(\tau)$ and $m_s(\tau)$ of a new independent variable τ that has the dimension of time as the Laplace inversions of functions $\hat{p}^r(\varpi), \hat{m}(\varpi)$ and $\hat{m}_s(\varpi)$ with respect to variable ϖ, that is,

equation (24)

$$\hat{p}^r = \int_0^\infty p^r(\tau)e^{-\varpi\tau}d\tau, \qquad \hat{m}_\rho = \int_0^\infty m_\rho(\tau)e^{-\varpi\tau}d\tau, \qquad \hat{m}_S = \int_0^\infty m_S(\tau)e^{-\varpi\tau}d\tau,$$

then the inverse Laplace transformation of the Eq. (22) for $\hat{p}^r$ with respect to the independent variable ϖ yields

equation (25)

$$\frac{d}{dt}p^r + \frac{\partial}{\partial\tau}p^r = m_\rho\frac{d\rho}{dt} + m_S\frac{ds}{dt},$$

where by definition $p^r(\tau \leq 0) = 0$.

Since

equation (26)

$$I = \int_0^\infty \hat{p}^r(\varpi)d\varpi = \int_0^\infty p^r(\tau)\int_0^\infty e^{-\varpi\tau}d\varpi d\tau = \int_0^\infty p^r(\tau)\tau^{-1}d\tau,$$

we obtain

equation (27)

$$p_\Sigma = p(\rho, s) + \int_0^\infty p^r(\tau)\tau^{-1}d\tau,$$

where I is the total relaxation pressure.

As a result, an infinite set of equations for the functions p_k^r has been reduced to a single equation of Boltzmann type (25) for the function p^r.

Eqs. (15 - (19), (25), (27) constitute the final system of governing equations for media with continuously distributed relaxation times which is being presented here for the first time.

Strictly speaking, appearance of phenomenological evolutionary Eq. (25) results in the requirement that the rate of entropy production must be a positive functional of p^r on the solutions of the governing system of equations. One way to work out the structure of the entropy equation in this situation is to use the formalism of the extended irreversible thermodynamics, whose detailed description can be found in the monographs [29] and [30], where the relationship between the conventional quasi-equilibrium irreversible thermodynamics and the extended irreversible thermodynamics are discussed in detail. The involved analytical procedure closely follows similar derivation for viscous heat conducting fluids described in Section 7.2 of monograph [29]and is not reproduced here for this reason. It results in the following extended Gibbs equation for the specific entropy:

equation (28)

$$T\frac{ds}{dt}=\left(\frac{de}{dt}+p\Sigma\frac{dv}{dt}\right)-\frac{1}{2\rho^2}\int_0^{\infty}(p^r)^2\frac{\partial}{\partial\tau}\left(\frac{1}{\tau m_\rho}\right)d\tau,$$

where the integral on the right-hand side of Eq. (28) must be negative. The sufficient (although not necessary) condition of this is $\partial(\tau m_\rho)/\partial\tau\geq 0$. It can be inferred from the analysis of the next section that the additional terms appearing in the entropy equation(28) have negligible higher-order effect on the properties of the acoustic field considered in this work.

System (15 - (19), (25), (27) contains phenomenological functions $m_\rho(\tau,\rho,s)$ and$m_s(\tau,\rho,s)$ which determine the absorption and dispersion properties of the medium. Our aim now is to obtain these functions in terms of the experimentally measurable absorption coefficient in the frequency domain. This will allow us to formulate a closed physical model for media with continuous distribution of relaxations.

4. DETERMINATION OF THE ABSORPTION FUNCTIONS

Assume that unsteady perturbations of the hydrodynamic parameters in the medium at rest are small, and consider a linear version of the governing equations. This can be obtained if a solution is sought in the form of the following series

equation (29)

$$\rho = \rho_o + \rho' + \cdots, \qquad T = T_o + T' \cdots, \qquad s = s_o + s' + \cdots,$$

equation (30)

$$p_\Sigma = p_o + p'_\Sigma + \cdots, \qquad \underline{v} = \underline{v}' + \cdots, \qquad p^r = p^r_r + \cdots,$$

equation (31)

$$p = p_o + c_{so}^2 \rho' + \frac{c_{po} - c_{vo}}{\beta_o c_{vo}} \rho_o s' + \cdots,$$

equation (32)

$$m_\rho = m_{\rho o}(\tau) + \cdots, \qquad m_s = m_{so}(\tau) + \cdots$$

where the subscript “o” refers to the unperturbed values of hydrodynamic parameters. All dashed variables represent the leading order corrections to the related unperturbed values.

Substitution of these expansions into the equations of motion , , , , (19) and (20), (25),(27) results in the following system

equation (33)

$$\frac{\partial^2 \rho'}{\partial t^2} - \nabla^2 p'_\Sigma - \frac{4\eta_o/3 + \varsigma_o}{\rho_o} \nabla^2 \frac{\partial \rho'}{\partial t} = 0,$$

equation (34)

$$\frac{\partial p_r'}{\partial t} + \frac{\partial p_r'}{\partial \tau} = m_{\rho o}\frac{\partial \rho'}{\partial t} + m_{so}\frac{\partial s'}{\partial t},$$

equation (35)

$$p_\Sigma' = c_{so}^2 \rho' + \frac{c_{po} - c_{vo}}{\beta_o c_{vo}} \rho_o s' + \int_0^\infty p_r' \frac{d\tau}{\tau}.$$

equation (36)

$$\frac{\partial T'}{\partial t} = \frac{\beta_o T_o c_{so}^2}{c_{vo} \rho_o} \frac{\partial \rho'}{\partial t} + \frac{\kappa_o}{c_{vo} \rho_o} \nabla^2 T',$$

equation (37)

$$\frac{\partial s'}{\partial t} = \frac{\kappa_o}{\rho_o T_o} \nabla^2 T'.$$

In the absence of viscosity, thermal conductivity and relaxations this system leads to conventional wave equation for the function ρ'.

In the presence of absorption the shape of the linear sinusoidal sound wave changes on a length scale proportional to the reciprocal of the absorption coefficient α_{ab}. As a result, there exist two characteristic length scales in lossy media—the wavelength and the absorption length equal to $1/\alpha_{ab}$. Their ratio defines the nondimensional absorption coefficient $\varepsilon = c_{so}\alpha_{ab}/f$, where f is the frequency of the sound wave.

In biological media this value is, normally, small. For example in [8], the amplitude attenuation coefficient for whole human liver *in-vivo* is given by the approximate formula $\alpha_{ab} \approx 0.052 f$ Np/cm, where the frequency of the sound wave f is measured in MHz. This gives the attenuation length of 19.2 cm, if the reference frequency f is taken to be 1 MHz. With the speed of sound in liver being $c_{so} \approx 1580$ cm/s, the value of the wavelength corresponding to $f = 1$ MHz is equal to $c_{so}/f \approx 1.58 \cdot 10^{-1}$ cm. Although the absorption coefficient is not exactly equal to the attenuation coefficient (but, normally, is somewhat smaller) and the actual values of the attenuation coefficient vary between different literature sources, it is important here that the value of the nondimensional absorption coefficient is a reasonably small quantity approximately equal to $\varepsilon=8.2\cdot10^{-3}$. Consequently, a typical value of the nondimensional absorption coefficient can be used as a natural small parameter ε for further asymptotic analysis of the system , , , and .

By virtue of the linear character of the problem, $m_{\rho o}(\tau)$ must be proportional to α_{ab}, and the coefficient $m_{\rho o}(\tau)$ can also be expected to be small.

The alternative way of introducing the small parameter ε is to define it as the ratio of some typical value of the function $m_{\rho o}$ to c_{so}^2, since, in agreement with formula (10), function $m_{\rho o}$ has the same dimension as the square of the speed of sound.

Thus, let

equation (38)

$$m_{\rho o} = \varepsilon\sigma(\tau), \qquad \eta_o = \varepsilon\eta', \qquad \zeta_o = \varepsilon\zeta', \qquad \kappa_o = \varepsilon\kappa',$$

where ε is a small parameter and $\sigma(\tau)$, η',ζ',κ' are assumed to be of the order of unity in ε.

Consider the problem of one-dimensional wave propagation in the direction of the positive x_1-axis. Assume that the wave field is excited by the sinusoidal motion of a piston which has started at $t=0$. The mean position of the surface of the piston coincides with the plane $x_1=0$.

The solution to the system, (33 – 37) can be constructed as an asymptotic expansion in ε:

equation (39)

$$\rho' = \rho_1(\theta, \xi) + o(1), \qquad p'_{\Sigma} = p_1(\theta, \xi) + o(1),$$

equation (40)

$$s' = \varepsilon s_1(\theta, \xi) + o(\varepsilon), \qquad T' = T_1(\theta, \xi) + o(1), \qquad p'_r = \varepsilon p_r(\theta, \xi, \tau) + o(\varepsilon),$$

where the independent variable $\theta = t - x_1/c_{so}$ describes fast changes of the phase of the wave whereas the independent variable $\xi = \varepsilon x_1$ describes slow variations of the amplitude and phase due to absorption and dispersion.

Substituting formulae (38) – (40) into the system (33 -37) retaining only the leading order terms in ε we obtain the following system of equations

equation (41)

$$2\frac{\partial p_1}{\partial \xi} = c_{so}^{-1} \frac{\partial}{\partial \theta} \int_0^\infty p_r \frac{d\tau}{\tau} + \lambda \frac{\partial^2 p_1}{\partial \theta^2},$$

equation (42)

$$\frac{\partial p_r}{\partial \theta} + \frac{\partial p_r}{\partial \tau} = \frac{\sigma(\tau)}{c_{so}^2} \frac{\partial p_1}{\partial \theta} + \mu \frac{\beta_o m_{so}(\tau)}{\rho_o} \frac{\partial^2 p_1}{\partial \theta^2},$$

equation (43)

$$\rho_1 = c_{so}^{-2} p_1, \qquad s_1 = \mu \frac{\beta_o}{\rho_o} \frac{\partial p_1}{\partial \theta}, \qquad T_1 = \frac{\beta_o T_o}{c_{po} \rho_o} p_1,$$

where

equation (44)

$$\lambda = \frac{1}{\rho_o c_{so}^3} \left(\zeta' + \frac{4}{3}\eta' + \kappa' \left(\frac{1}{c_{vo}} - \frac{1}{c_{po}} \right) \right), \qquad \mu = \frac{\kappa'}{c_{po} \rho_o c_{so}^2}.$$

The initial and boundary conditions are

equation (45)

$p_1(\theta \le 0) = p_r(\theta \le 0) = p_r(\tau \le 0) = 0,$

equation (46)

$$p_1(\xi = 0) = A\sin(\omega\theta), \qquad p_1(\theta, \xi \to \infty) \to 0,$$

where $A>0$ is a given amplitude of pressure perturbation at $\xi=0$.

The solution to this problem can be obtained by means of the Laplace transformation with respect to θ, thus

equation (47)

$$p_1 = \frac{A\omega}{2\pi i}\int_{\gamma'-i\infty}^{\gamma'+i\infty} \varphi(u)\exp(u\theta)du,$$

where $\gamma'>0$,

equation (48)

$\varphi(u) = \exp(uc(u)\xi/2)/(\omega^2+u^2),$

equation (49)

$$c(u) = c_{so}^{-1}\int_0^\infty \int_0^\tau \left\{ u\frac{\sigma(\tau')}{c_{so}^2} + \mu u^2 \frac{\beta_o m_{so}(\tau')}{\rho_o} \right\} e^{-u(\tau-\tau')}d\tau' \frac{d\tau}{\tau} + \lambda u.$$

At this stage of analysis the function c(u) remains unknown, since our aim is to link$\sigma(\tau)$ and $m_{so}(\tau)$ to the absorption coefficient of the medium $\alpha_{ab}(\omega)$.

It can be expected on physical grounds that after a sufficiently long time, the initial perturbations created when switching the piston on will have decayed, with the remaining portion of the sound field being entirely due to the forced periodic motion of the piston. For this reason, the knowledge of the transient part of the solution (47) is not needed to determine the functional relationship between functions $\sigma(\tau)$ and $m_{so}(\tau)$ and the absorption coefficient $\alpha_{ab}(\omega)$. For this purpose we expand the solution (47) at long retarded time θ.

In order to obtain this asymptotic expansion, additional assumptions have to be imposed on the structure of the function $\varphi(u)$. These can be formulated on the basis of Theorem 41.1 in [31] and the discussion therein. We require that the function $\varphi(u)$ of the complex variable u (i) be analytic in the half-plane $\mathrm{Re}(u) > \gamma$; (ii) have any number of poles of any order or branching points in the left half-plane $\mathrm{Re}(u) < 0$, and (iii) have any finite number of branching points or zeros $\alpha_1,\alpha_2,\alpha_3,\ldots,\alpha_{n-2}$ on the vertical line $\mathrm{Re}(u) = 0$, but no order one poles, apart from the poles and $\alpha_n = i\omega$ which correspond to the driving frequency of the piston, *i.e.* , in the vicinity of any point α_j, where j=1,2,3,…,n−2, the function $\varphi(u)$ can be expanded into the series

equation (50)

$$\varphi(u) = \sum_{\nu=0}^{\infty} c_\nu (u - \alpha_j)^{\lambda_\nu^j}, \quad -1 < \lambda_0^j < \lambda_1^j < \cdots < \infty.$$

According to the results formulated in [31] and also in [32], the asymptotic expansion of the inverse Laplace transform (47) as θ goes to plus infinity is then given by the term

equation (51)

$$p_1 = Ae^{-0.5\omega c_i(\omega)\xi} \sin\left(\omega\left(\theta + \frac{1}{2}c_r(\omega)\xi\right)\right) + \cdots,$$

plus the sum of functions that describe the decaying transient motion. Here $c_r(\omega) + ic_i(\omega) = c(i\omega)$,

equation (52)

$$c_r(\omega) = \frac{\omega}{c_{so}} \int_0^{\infty} \int_0^{\tau} \frac{\sigma(\tau')}{c_{so}^2} \sin \omega(\tau - \tau') - \mu\omega \frac{\beta_o m_{so}(\tau')}{\rho_o} \cos \omega(\tau - \tau') d\tau' \frac{d\tau}{\tau},$$

equation (53)

$$c_i(\omega) = \frac{\omega}{c_{so}} \int_0^{\infty} \int_0^{\tau} \frac{\sigma(\tau')}{c_{so}^2} \cos \omega(\tau - \tau') + \mu\omega \frac{\beta_o m_{so}(\tau')}{\rho_o} \sin \omega(\tau - \tau') d\tau' \frac{d\tau}{\tau} + \lambda\omega,$$

where τ' is the integration variable.

In agreement with formula (51), the absorption coefficient is given by the following expression

equation (54)

$$\alpha_{ab}(\omega) = \varepsilon\omega c_i(\omega)/2.$$

After substitution of formula (54) for $c_i(\omega)$, and formula (44) for λ, Eq. (54) takes the form

equation (55)

$$\alpha_{ab}(\omega) = \frac{\varepsilon\omega^2}{2c_{so}} \int_0^{\infty} \int_0^{\tau} \left\{ \frac{\sigma(\tau')}{c_{so}^2} \cos \omega(\tau - \tau') + \mu\omega \frac{\beta_o m_{so}(\tau')}{\rho_o} \sin \omega(\tau - \tau') \right\} d\tau' \frac{d\tau}{\tau} + \alpha_{ab}^{n}(\omega).$$

where

equation (56)

$$\alpha_{ab}^{n}(\omega) = \frac{\varepsilon\omega^2}{2\rho_o c_{so}^3} \left(\zeta' + \frac{4}{3}\eta' + \kappa' \left(\frac{1}{c_{vo}} - \frac{1}{c_{po}} \right) \right)$$

is the well known expression for the absorption coefficient in a viscous heat conducting Newtonian fluid, [2]. This is of no interest for the subsequent study, and in what follows only the relaxation part of the absorption coefficient

equation (57)

$$\alpha^{r}_{ab}(\omega) = \alpha_{ab}(\omega) - \alpha^{n}_{ab}(\omega)$$

is considered.

The nondimensional combination of parameters

equation (58)

$$\mu\omega = \frac{\kappa'\omega}{c_{po}\rho_o c^2_{so}}$$

in formula (55) characterizes the influence of the thermal conductivity on the relaxation part of the absorption coefficient $\alpha^{r}_{ab}(\omega)$. In many practical situations this value is small, and the thermal conductivity is negligible. This is certainly true for most biological applications. For example, the value of μω based on the experimental data presented in [8] for the typical frequency of 1 MHz results in the assessment $\mu\omega \sim 10^{-6}$ for most soft tissues and biological fluids. For this reason, the term containing the coefficientμω will be neglected.

In terms of functions $\alpha^{r}_{ab}(\omega)$ and $m_{po}=\varepsilon\sigma(\tau)$ given by formulae (38) and (57), respectively, formula (55) can then be rewritten as the following integral equation for the unknown rheological function m_{po}:

equation (59)

$$\int_0^\infty \int_0^\tau m_{po}(\tau')\cos\omega(\tau-\tau')d\tau'\frac{d\tau}{\tau} = \frac{2c^3_{so}}{\omega^2}\alpha^{r}_{ab}(\omega).$$

4.1. Solution of the integral equation for the absorption function $m_\rho(\tau)$

Our aim now is to obtain an analytic formula for $m_{\rho o}(\tau)$ as a function of absorption coefficient $\alpha^r_{ab}(\omega)$, in order to formulate a closed system of equations of motion for media whose absorption properties are described by the absorption coefficient $\alpha^r_{ab}(\omega)$ as a function of the sound frequency $\omega \geq 0$.

If we change the order of integration in the left-hand side of formula (59), make the substitution $y=\tau-\tau'$ in the resulting inner integral and then change the order of integration to the original one, after the introduction of the function

equation (60)

$$\psi(y) = \int_0^\infty m_{\rho o}(\tau')\frac{d\tau'}{y+\tau'}$$

the integral equation (59) takes the form

equation (61)

$$\int_0^\infty \psi(y')\cos\omega y'\, dy' = f(\omega),$$

where $f(\omega) = 2c^3_{so}\omega^{-2}\alpha^r_{ab}(\omega)$ and $0 \leq \omega < +\infty$.

Multiplication of the left-hand side of Eq. (61) by $\cos\omega y\, d\omega$, where $y>0$, and integration over the interval $0 \leq \omega < +\infty$ yields

equation (62)

$$\psi(y) = \frac{2}{\pi}\int_0^\infty f(\omega)\cos\omega y\, d\omega.$$

Now consider analytic continuation of (60) to the right half-plane $\mathrm{Re}(y) > 0$. By the residue theorem, the inverse Laplace transformation of the right-hand side gives the relationship

equation (63)

$$\mathcal{L}_u^{-1}\{\psi(y)\} = \int_0^{\infty} m_{\rho o}(\tau) e^{-u\tau} d\tau = \hat{m}_{\rho o}(u),$$

i.e. , the inverse Laplace transform of $\psi(y)$ is equal to the Laplace transform $\hat{m}_{\rho o}(u)$ of the function $m_{\rho o}(\tau)$.

In view of formula (62), the unknown function $m_{\rho o}(\tau)$ can be rewritten in terms of the absorption coefficient as

equation (64)

$$m_{\rho o}(\tau) = \frac{4c_{so}^3}{\pi}\left(\frac{1}{2\pi i}\right)^2 \int_{\gamma_1 - i\infty}^{\gamma_1 + i\infty} e^{u\tau} du \int_{\gamma_2 - i\infty}^{\gamma_2 + i\infty} e^{uy} dy \int_0^{\infty} \frac{\alpha_{ab}^{r}(\omega)\cos\omega y}{\omega^2} d\omega,$$

where $\gamma_1, \gamma_2 \geq 0$.

Formula (64) solves the integral equation (59) and gives the required representation of the unknown phenomenological function $m_{\rho o}(\tau)$ versus the experimentally measurable value of the absorption coefficient $\alpha_{ab}^{r}(\omega)$. Together with the relaxation equation (25) and the equation of state (27) it solves the problem of reconstruction of the equations of motion from experimental measurements.

The advantage of these equations over previous approaches to this problem that have been reported in the literature (see Introduction for more details) is that these formulae are not restricted to any specific form of the functional dependency of the absorption coefficient on the frequency ω, such as the power law absorption, for instance.

The only essential physical restrictions on applicability of formula (64) is the well-proven hypothesis of negligible thermal conductivity on the time scale of acoustic motion, and the finite value of the total relaxation pressure on the right-hand side of formula (27).

5. EXAMPLES OF APPLICATIONS

The adiabatic version of theory (s=μ=0), proposed in the previous section, will now be applied to a number of experimental laws of sound absorption used in different applications. The following examples demonstrate the use of formula (64), in order to recover the function $m_\rho(\tau)$ which appears in the relaxation equation (25) and characterizes the absorption and dispersion properties of the medium.

A knowledge of the rheological function $m_\rho(\tau)$ is important because it gives a closed set of equations which describes the wave propagation process. This new set of equations consists of the standard hydrodynamic equations plus an additional equation for the new state variable p^r which appears in the thermodynamic equation of state, and introduces coupling with the internal structure of the material. Further theoretical discussion of this approach can be found in [24] and references therein.

A full nonlinear version of the resulting governing system is given by (15-20), (25), (27). Its linear counterpart for uniform media is the system (33-37).

In many practical situations the non-dimensionalized values of the acoustic velocity $\underline{v}$, the density derivative $d\rho/dt$ and the absorption function m_ρ are small. Since the relaxation pressure p^r is of the order of the product $m_\rho d\rho/dt$, the convective term $\underline{v}\cdot\nabla\hat{p}^r$ in(22) is $o(\underline{v}^2)$ and can often be neglected. This is certainly true for a quadratically nonlinear theory. In this case the linear version of the Eq. (22) can be used. Its solution is given by the following expression

equation (65)

$$\hat{p}^r = \int_0^t \frac{d\rho}{dt'}\hat{m}_\rho(\varpi)\exp(-\varpi(t-t'))dt'.$$

which after the substitution into the first equality (26) gives the following formula for the total relaxation pressure

equation (66)

$$I = \int_0^t \frac{d\rho}{dt'} J_i(t - t')dt',$$

where i=1,2 and

equation (67)

$$J_1(y) = \int_0^\infty \frac{m_\rho(\tau)d\tau}{\tau + y} \quad \text{or} \quad J_2(y) = \frac{4}{\pi} c_s^3 \int_0^\infty \frac{\alpha^r_{ab}(\omega)\cos\omega y}{\omega^2} d\omega,$$

if use of formula (64) is made. Since $p_\Sigma = p(\rho,s) + I$, the nonlinear momentum equation then takes the following compact form:

equation (68)

$$\frac{dv}{dt} + \frac{1}{\rho}\nabla p(\rho,s) - \frac{1}{\rho}\nabla \int_0^t t[\rho\nabla \cdot v](t')J(t - t')dt' = 0,$$

where $J(y)=J_1$ or $J(y)=J_2$. It must be supplemented with the continuity Eq. (15) and the equation of state

equation (69)

$$p(\rho,s) = \text{const} + c_{so}^2(\rho - \rho_o) + \frac{B}{2}(\rho - \rho_o)^2 + \cdots,$$

where B is the nonlinearity parameter of the medium, and $\rho_o(r)$ is the distribution of density in the medium at rest.

A number of examples of how the wave equations in different lossy media can be generated with the help of this system is given in this section. The methods for their numerical solution comprise a separate topic which goes beyond the scope of this paper and will be addressed elsewhere.

5.1. Example 1: finite number of relaxations

This example demonstrates how the case of a medium with a finite number of relaxation frequencies can be derived from the present theory with continuously distributed relaxations proposed in the previous sections. A finite number of relaxations is typical, say, for atmospheric absorption, where the absorption coefficient is the sum of two relaxation terms associated with the vibration of nitrogen and oxygen molecules plus the Newtonian term (56), [3]:

equation (70)

$$\alpha_{ab}(\omega) = \alpha^{n}_{ab}(\omega) + \sum_{n} \frac{\varepsilon}{2} \frac{m_n}{\varpi_n} \frac{\omega^2}{1 + \omega^2 \varpi_n^{-2}},$$

where m_n are constant, $\varpi_n > 0$ are the discrete relaxation frequencies of the medium, and n is an integer.

If the expression for $\alpha^{r}_{ab}(\omega) = \alpha_{ab}(\omega) - \alpha^{n}_{ab}(\omega)$, where $\alpha_{ab}(\omega)$ is defined by (70), is substituted into the second formula (67), it can be shown after some algebra that

equation (71)

$$J(y) = \varepsilon \sum_{n} m_n e^{-y\varpi_n}.$$

Hence, the final expression for the total relaxation pressure is:

equation (72)

$$I = \varepsilon c^{3}_{so} \sum_{n} m_n \int_{0}^{t} \frac{d\rho}{dt'} e^{-\varpi_n (t-t')} dt'$$

which coincides with the well known formula for multiple discrete relaxations, [2].

5.2. Example 2: absorption of sound in soft tissue and biological fluids

For many types of soft tissue and biological fluids the attenuation coefficient is given by the empirical formula, [8]:

equation (73)

$$\alpha^{r}_{ab}(\omega) = \varepsilon\alpha_{o}\omega^{1+r},$$

known as a power law of absorption. Here $\alpha_o>0$ and $0<r<1$ are experimental constants.

According to the second formula (67), the power law of absorption (73) implies that

equation (74)

$$J(y) = \frac{4}{\pi}c^{3}_{so}\varepsilon\alpha_{o}\Gamma(r)\cos(\pi r/2)y^{-r}.$$

Consequently, the total relaxation pressure (66) is the fractional derivative of the density

equation (75)

$$I = \frac{4\cos(\pi r/2)\Gamma(r)}{\pi}\varepsilon\alpha_{o}c^{3}_{so}\int_{0}^{t}\frac{d\rho}{dt'}\frac{dt'}{(t-t')^{r}}.$$

If r=0 (*i.e.* , when $\alpha_{ab}(\omega)$ is linear in ω), then Γ(r)=∞ and the right-hand is unbounded.

Calculation of the absorption function results in the expression

equation (76)

$$m_{\rho o}(\tau) = \frac{4\cos(\pi r/2)\sin(\pi r)\Gamma(r)}{\pi^{2}}\begin{cases}0, & \text{if } \tau \leq 0\\ \varepsilon\alpha_{o}c^{3}_{so}\tau^{-r}, & \text{if } \tau > 0.\end{cases}$$

The unidirectional 1D linear problem , , , (44), and for the pressure field p_1 then reduces to an integro-differential equation of the Volterra type:

equation (77)

$$\partial_\xi p_1 = \frac{2\cos(\pi r/2)\Gamma(r)}{\pi}\alpha_o\partial_\theta \int_0^\theta \frac{\partial p_1(\theta')}{\partial \theta'}\frac{d\theta'}{(\theta-\theta')^r}$$

with the related initial and boundary conditions. Here, as elsewhere in the paper, and for consistency with other literature sources mentioned in the Introduction, we discard the Newtonian absorption term and set $\lambda=0$.

This equation is also known in the literature as the equation of fractional dynamics, [33]. An in depth discussion of the fundamental issues of derivation of such equations can be found in [34]. In recent years this topic has received a great deal of attention in several acoustics publications, such as [9], [35] and [11], in which the absorption mechanism was modelled in terms of fractional derivatives. In this section we have demonstrated how this approach can be derived as a special case of the theory with a continuous distribution of relaxation processes.

The solution to the unidirectional equation (77) is

equation (78)

$$p_1 = A\omega \mathcal{L}_u^{-1}\left\{(\omega^2+u^2)^{-1}\exp\left(u\theta + \frac{\alpha_o}{\sin \pi r/2}u^{r+1}\xi\right)\right\}.$$

The leading term of the asymptotic expansion of integrals of this type when θ goes to infinity is determined by the poles of the integrand, [32], that is,

equation (79)

$$p_1 \approx A\exp(-\alpha_o\omega^{r+1}\xi)\sin(\omega\theta+\cot(\pi r/2)\alpha_o\omega^{r+1}\xi).$$

As expected, this last formula gives the same absorption coefficient $\alpha^{r}_{ab}(\omega)$ as that quoted at the beginning of this subsection. The increase of the speed of sound over the speed of sound in the medium at rest given by this formula is

equation (80)

$$\Delta c_s = \cot(\pi r/2)\varepsilon\alpha_o\omega^{r+1}.$$

This result is consistent with the related dispersion relation obtained in [36] from formula (73) using the causality argument based on the regularized Kramers–Kronig relation.

The related non-unidirectional linear equation of fractional dynamics considered in [12] is readily recovered on linearization of the system (15), (68), (69), with the function J being defined by formula (74).

5.3. Example 3: linear law of absorption

Consider now the linear law of absorption

equation (81)

$$\alpha^{r}_{ab}(\omega) = \varepsilon\alpha_o\omega.$$

Formula (81) gives a reasonably good fit to experimental data in the MHz frequency range for soft tissue and some biological fluids, [8], and, owing to its simple structure, plays an important role in theoretical studies, [37].

Formula (81) is the special case of the power law of absorption (73) with r=0. In this case the system of governing equations becomes singular, since $\Gamma(r=+0)=+\infty$ in (74).

Formula (80) also shows that in the limit of linear absorption the solution obtained in the previous subsection for the power law of absorption develops a singularity. Indeed, Δc_s blows up as r goes to zero.

In order to overcome this difficulty, the authors of [36] proposed a form of regularized Kramers–Kronig relation specifically tailored for the case of the linear absorption law(81). In the limit r=+0 their dispersion relation (*cf.* formula (88)) results in the logarithmic dependence of Δc_s on ω, rather than expression (80). However, the exact structure of the governing equation of motion in physical time-space has remained unclear.

The approach we propose in this paper is particularly suitable for answering this question. It also sheds extra light on the mathematical and physical nature of the singularity.

In order to illustrate the essence of the problem and to recover the system of governing equations which corresponds to the regularized Kramers–Kronig relation used in [36] to treat the case of linear absorption law, consider the function $m_{po}(\tau)=\varepsilon\sigma(\tau)$, where $\sigma(\tau)$ is taken in the form

equation (82)

$$\sigma(\tau) = \alpha_o c_{so}^3 \begin{cases} 0, & \text{if } \tau < 0 \\ \exp(-\Delta\tau), & \text{if } \tau > 0 \end{cases}$$

where Δ>0 is a constant. Then, the first formula (67) results in the expression

equation (83)

$$J(y) = \frac{4}{\pi} c_{so}^3 \varepsilon \alpha_o e^{\Delta y} \int_{\Delta y}^{\infty} \frac{e^{-z} dz}{z},$$

with the related nonlinear governing system given by the Eqs. (15), (68), (69). Its linear unidirectional 1D version (41), (46) reduces to the following integro-differential equation:

equation (84)

$$\partial_\xi p_1 = 2\alpha_o \int_0^\theta \frac{\partial}{\partial\theta'} p_1(\theta') \int_0^\infty \frac{e^{-\Delta z}}{z+\theta-\theta'} dz d\theta'.$$

Solution of Eq. (84) by means of the Laplace transformation in θ and subsequent evaluation of the asymptotic behaviour of the function p_1 for large θ results in the following formula

equation (85)

$$p_1(\theta \to +\infty) = \exp(-\alpha_{ab}^{r}(\omega)\xi)\sin\left(\omega\theta - \frac{0.5\alpha_o\omega^2}{\Delta^2+\omega^2}\xi\left(\omega \ln\left(\frac{\Delta}{\omega}\right) - \Delta\frac{\pi}{2}\right)\right).$$

where the absorption coefficient is

equation (86)

$$\alpha_{ab}^{r}(\omega) = \frac{0.5\varepsilon\alpha_o\omega^2}{\Delta^2+\omega^2}\left(\omega\frac{\pi}{2} - \Delta \ln\left(\frac{\omega}{\Delta}\right)\right).$$

For low frequencies (ω→0) this gives the slightly higher than quadratic growth of absorption with frequency observed in Newtonian fluids.

In the limit of small Δ the solution is a pressure wave with the absorption coefficient linearly dependent on the frequency:

equation (87)

$$p_1(\theta \to \infty) = e^{-\frac{\pi}{4}\alpha_o\omega\xi} \sin\left(\omega\theta + \frac{\alpha_o\omega}{2} \ln\left(\frac{\omega}{\Delta}\right)\xi\right) + \cdots.$$

This formula gives exactly the same dispersion relation as the dispersion relation given as Eq. (13) in [36] derived from the regularized Kramers–Kronig relation for the case of linear absorption law (81):

equation (88)

$$\frac{1}{c_{\mathrm{rel}}(\omega)} - \frac{1}{c_{so}} = -\frac{\alpha_o \omega}{2} \ln\left(\frac{\omega}{\Delta}\right),$$

where $c_{\mathrm{rel}}(\omega)$ is the phase speed at frequency ω in the medium with relaxations. It is essential that the parameter Δ is identical to the reference frequency ω_o of [36]. Consequently, we see that this reference frequency is not arbitrary, but has a precise physical meaning. It determines the rate of decay of the rheological function (82) at infinity and the slope of the absorption coefficient (86) at low frequencies in the law of absorption (86) which correspond to the regularized dispersion relation (88).

5.4. Combined absorption function for biological tissues

Although the power law of absorption (73) fits well with experimental data in the MHz frequency range, at low and very high frequencies it may not be appropriate, [8].

An interesting two parameter absorption model

equation (89)

$$\alpha^{\mathrm{r}}_{\mathrm{ab}}(\omega) = \varepsilon\alpha_* \omega \arctan \beta_* \omega, \qquad \alpha_*, \beta_* = \mathrm{const}$$

which combines the classical quadratic law in the lower frequency limit with the linear law of absorption in the MHz frequency range was proposed in [38] for biological tissues. According to [38], this model is almost an order of magnitude more accurate than the power law of absorption. The major improvement of formula (89) over the linear formula (81) is in the lower frequency range.

In agreement with the second formula (67), the function J(y) for the absorption law (89)is given by the formula

equation (90)

$$J(y) = 2c_{so}^3 \varepsilon \alpha_o \int_{y/\beta}^{\infty} \frac{e^{-z} dz}{z},$$

with the related governing equations being (15), (68), (69).

Fig. 1 presents the comparison of absorption models for the function α_{ab}^r based on the formulae (70), (73), (86), (89) for the case of homogenized bovine liver studied experimentally in [38].

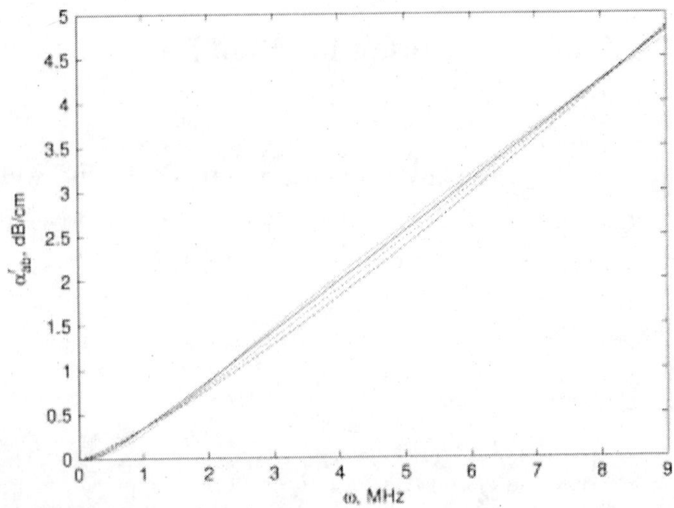

Figure. 1. Absorption coefficient α_{ab}^r versus frequency ω: solid line–combined formula (89) obtained experimentally in [38], dashed line–power law (73), short dashed line–formula (86), dotted line–two discrete relaxations(70). In agreement with formulae (70), (73), (86), (89), in all cases $\alpha_{ab}^r(\omega = 0) = 0$.

Following [38], we take $\varepsilon\alpha_* = 2\pi \times 0.057\,\mathrm{dB/(cm\,MHz)}$, $\beta_* = 2\pi \times 0.22\,1/\mathrm{MHz}$ in formula (89)and $\varepsilon\alpha_o = 0.34\,\mathrm{dB/cm}, r = 0.21$ in the power law formula (73), respectively. According to Jongen et al. [38], the minimal least square deviations from the power law (73), and the combined absorption function (89) from the experimental data obtained by these authors in the range between 1 and 9 MHz are equal to 0.047 and 0.008, respectively. Jongen et al. report that the agreement between their measurements made, for samples obtained from 7 livers, is within 5%. Further details concerning the accuracy of the above experimental data and comparisons with the fits provided by formulae (73) and (89) can be found in [38].

Unfortunately, the absence of tabulated experimental data in [38] precludes making similar fits and related mean square error assessments for the absorption laws(70) and (86). For this reason, the unknown parameters in the absorption laws(70) and (86) were chosen so as to insure the best agreement with the experimental formula (89) in the frequency range $0 < \omega < 9\,\mathrm{MHz}$, for which the experiments were conducted in [38]. These are as follows: $\varepsilon\alpha_o = 0.8\,\mathrm{dB/cm}, \Delta = 0.82\,\mathrm{MHz}$ for the formula (86)and $n = 1, 2, \varepsilon m_1 = 1.25\,1/\mathrm{MHz}, \varepsilon m_2 = 1.5\,1/\mathrm{MHz}, \varpi_1 = 2.5\,\mathrm{MHz}, \varpi_2 = 11.5\,\mathrm{MHz}$ for formula (70), respectively. It can be seen from Fig. 1 that the overall quality of approximations based on the formulae (70) and (86) is at least as good as that provided by the experimentally obtained power law formula (73).

Our aim now is to compare the implications for the solutions of the lossy wave Eqs. (15),(68), (69) which follow from the four absorption laws shown in Fig. 1. This can be done by numerical solution of the linear and nonlinear versions of this system of equations, with the functions J(y) being given by formulae (71), (74), (83) and (90), respectively. Since the objective of this work is to study the properties of the absorption operators, only the one-dimensional version of the governing equations is considered, in order to exclude spatial effects. For the sake of simplicity, in our numerical study we neglect all nonlinear terms except the term

$0.5B(\rho-\rho_o)^2$ in the equation of state (69). Normally, in soft biological tissues this term dominates all other nonlinearities, owing to the large value of the coefficient of nonlinearity B. The following parameters of the medium are used in the computations, [8]: $\rho_o = 1070\ \mathrm{kg/m^3}$, $c_{so} = 1600\ \mathrm{m/s}$, $B = 7.5c_{so}^2/\rho_o$.

The medium occupies the interval $0 < x_1 < x_{\max}$ and is set into motion by the sinusoidal perturbation of the velocity on its left boundary $x_1=0$ with frequency $\omega = 1\ \mathrm{MHz}$. The initial and boundary conditions are: $v_1=0, \rho=\rho_o$ at $t=0$, $v_1(x_1=0)=A\sin(\omega t)$ and $v_1(x_1 = x_{\max}) = 0$. The position of the solid wall $x_1 = x_{\max}$ which bounds the computational domain on the right is taken far enough to ensure the absence of the reflected wave pattern in the bulk of the computational domain for sufficiently large times. To obtain an unbiased comparison between different lossy operators, no flux or slope limiters are implemented in the nonlinear numerical scheme in order to ensure computational stability on the shock fronts. Instead, this is achieved solely through the increase of the number of mesh points till the shock structure can be properly resolved.

Fig. 2 illustrates the behaviour of the sound pressure versus the propagation distance obtained from the solution of the linearized system of governing equations for the above four absorption laws. It can be inferred from Fig. 2 that the solutions based on the combined formula (89), formula (86) and the power law (73) are in a reasonably good agreement with each other, although the power law is in a somewhat worse agreement with the first two of these solutions. The solution based on the formula with two discrete relaxations (70) works well only within the distance equal to one wave period from the source, but does not give good agreement with the other three absorption models elsewhere.

Fig. 3 demonstrates the behaviour of the nonlinear solutions corresponding to the above four absorption laws. There is good agreement between pressure amplitude predictions based on the combined formula (89) and formula (86). The power law (73) and the formula with two discrete relaxations (70) result in higher pressure amplitudes and feature shock wave formation which do not occur in the solutions corresponding to formulae (89) and (86). Compared to the combined formula (89), at least a twelvefold increase in the number of mesh points per wavelength was needed to suppress numerical oscillations near the shock fronts, if formula (70) with two discrete relaxations was used. For the power law formula (73) an equivalent fivefold increase was required. However, the power law solution features large phase differences relative to the combined formula and the formula with two discrete relaxations which are in phase with each other.

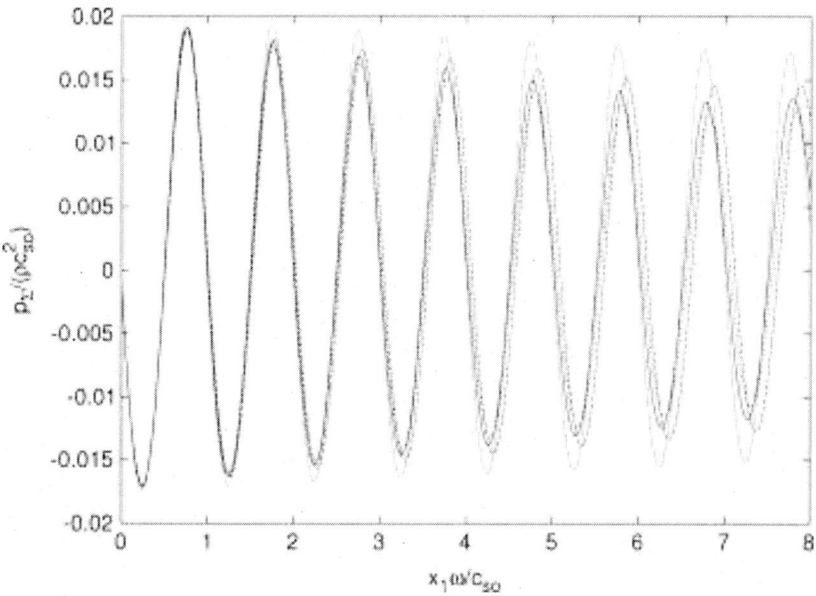

Figure. 2. Solution of the linear system of equations. Nondimensional sound pressure $p_\Sigma/(c_{s0}^2\rho_0)$ versus nondimensional distance $x_1\omega/c_{so}$ at time $t = 50.26/\omega$, $\omega = 1$ MHz: solid line–combined formula (89), dashed line–power law (73), short dashed line–formula (86), dotted line–two discrete relaxations (70).

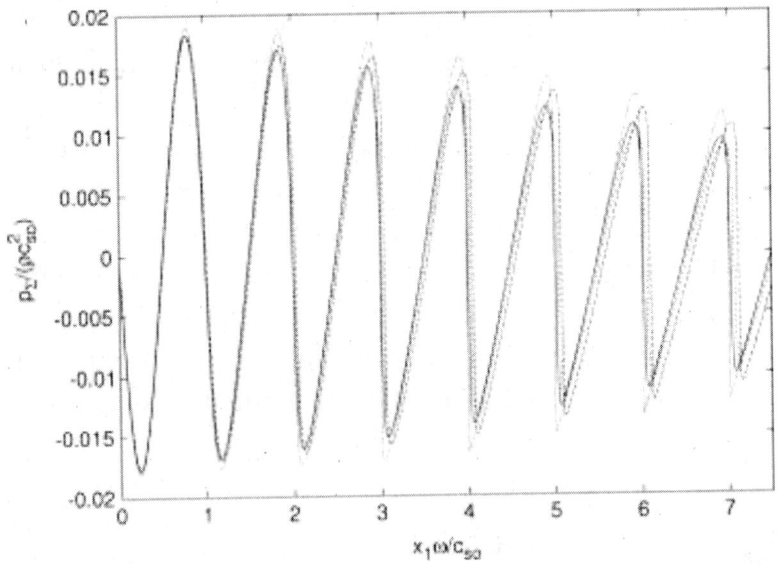

Figure. 3. Solution of the nonlinear system of equations. Nondimensional sound pressure $p_\Sigma/(c_{so}^2\rho_o)$ versus nondimensional distance $x_1\omega/c_{so}$ at time $t = 50.26/\omega$, $\omega = 1$ MHz: solid line–combined formula (89), dashed line–power law (73), short dashed line–formula (86), dotted line–two discrete relaxations (70).

Fig. 4 features the effects of nonlinearity on the solution. It depicts the sound pressure plots presented in Fig. 2 and Fig. 3 for the case of the combined absorption law (89). Nonlinear steepening of the wave profiles can be clearly observed. Also visible is the effect of strong nonlinear damping, as would be expected *a priori* from general theoretical considerations [2].

Overall, we see that the proposed methodology makes it possible to derive a complete set of equations for modelling the mechanism of absorption in tissue. In terms of numerical modelling, and in the context of HIFU applications, it makes sense to also employ models (86) and (89) described in Sections 5.3 and 5.4, along with the power law of absorption and the corresponding fractional wave equation. This is especially true bearing in mind that, over a wide range of frequencies, formula (89) is excellent agreement with the experimental data, [38]. Our experience indicates that the combined formula (89) can be computationally more favourable for the numerical treatment of acoustic problems with strong nonlinearity.

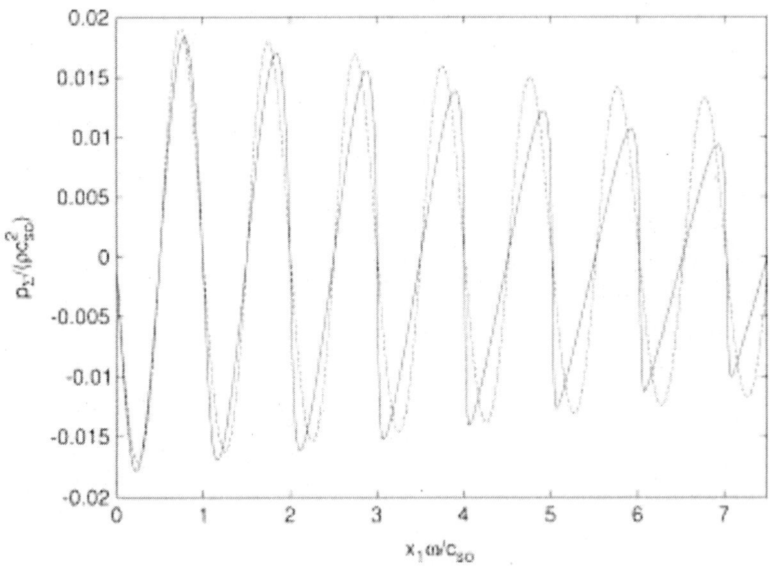

Figure. 4. Comparison of linear and nonlinear solutions based on the combined formula (89). Nondimensional sound pressure $p_\Sigma/(c_{so}^2\rho_o)$ versus nondimensional distance $x_1\omega/c_{so}$ at time $t = 50.26/\omega$, $\omega = 1$ MHz: solid line–nonlinear solution, dashed line–linear solution.

6. CONCLUSIONS

Accurate theoretical description of acoustic waves in fluids with complex internal structure is not normally possible within the framework of Newtonian fluid dynamics because this does not provide an adequate model of the absorption mechanism. Since the pioneering work of [16], it has been recognized that an extension of classical hydrodynamics incorporating the interaction between the macroscopic motion of the fluid and its internal structure is required in order to overcome this difficulty.

This can often be done by the introduction of a finite number of additional parameters characterizing the internal state of the liquid. This must be accompanied by the related modification of the governing hydrodynamic equations, in order to accommodate the additional set of dynamic equations for these new variables in a self-consistent way, [24].

However, in media such as biological fluids the situation is more complex. Since 1970, it has been realized that a finite number of internal variables is not sufficient to account for the properties of absorption in biological media, [6]. Although the concept of a continuous distribution of relaxation degrees of freedom was put forward, over the next forty years no relaxation theory with continuous distribution of relaxation frequencies had been formulated in acoustics.

The main result of this work is that it bridges this theoretical gap. More specifically, the coupled system of governing equations (15-20), (25), (27), and their linear counterpart (33-37), which link acoustic motion with the dynamic reaction of the internal structure of the fluid to this motion has been formulated in Section 3.

In the important case of adiabatic motion, when this system contains only one unknown rheological function $m_\rho(\tau,\rho,s)$, explicit expressions for this function in terms of the experimentally measurable absorption coefficient as a function of the wave frequency have been derived in Section 4. As a result, a closed system of equations of motion can be formulated directly from absorption measurement data on the basis of the proposed theory.

As demonstrated in Section 5, the theory presented here incorporates, as its special cases, the classical theory of discrete relaxations and the fractional wave equations recently proposed for biological tissues, [9], [11], [12] and [13].

The work presented clarifies the physical nature of the singularity which develops in the fractional wave equations in the case of linear absorption and provides a method for reconstruction of the wave equation corresponding to the regularization procedure described in the literature, [36], in order to remove this singularity.

One of the major methodological differences between the present theory and the fractal ladder approach of [13], is that the use of a discrete set of internal parameters in [13]results in the fact that their theory does not allow for continuous variation of the exponent(2) in the power law absorption. The present theory is free from this restriction. It also provides a much wider range of absorption laws than a power law. This feature of the present theory gives a much needed flexibility in the numerical treatment of the problem as has been illustrated by the examples presented in Section 5.

ACKNOWLEDGEMENT

The authors gratefully wish to acknowledge the support of the Engineering & Physical Sciences Research Council (EPSRC) via grant number EP/F0257501/1.

REFERENCES

1. A.D. Pierce, Acoustics. An Introduction to its Physical Principles and Applications, in: Sunnyside Boulevard, vol. 500, Published by the Acoustical Society of America, Woodbury, New York, 1991, p. 11797.

2. O.V. Rudenko, S.I. Soluyan, Theoretical Foundations of Nonlinear Acoustics, Consultants Bureau, NY, London, 1977.

3. T.D. Rossing (Ed.), Springer Handbook on Acoustics, Springer, 2007.

4. A.I. Nachman, J.F. Smith, R.C. Waag, An equation for acoustic propagation in inhomogeneous media with relaxation losses, J. Acoust. Soc. Am. 88 (3) (1990) 1584–1595.

5. P.-K. Choi, J.-R. Bae, K. Takagi, Ultrasonic spectroscopy in bovine serum albumin solutions, J. Acoust. Soc. Am. 87 (2) (1990) 874–881.

6. H. Pauly, H.P. Schwan, Mechanism of absorption of ultrasound in liver tissue, J. Acoust. Soc. Am. 50 (2B) (1971) 692–699.

7. C.R. Hill, J.C. Bamber, G.R. ter Haar (Eds.), Physical Principles of Medical Ultrasonics, second ed., John Wiley & Sons, Ltd., 2004

8. F.A. Duck, Physical Properties of Tissue. A Comprehensive Reference Book, Academic Press, 1990.

9. T.L. Szabo, Time domain wave equations for lossy media obeying a frequency power law, J. Acoust. Soc. Am. 96 (1) (1994) 491–500.

10. W. Chen, S. Holm, Modified Szabo's wave equations for lossy media obeying a frequency power law, J. Acoust. Soc. Am. 114 (5) (2003) 2570–2574.

11. W. Chen, S. Holm, Fractional Laplacian time-space models for linear and nonlinear lossy media exhibiting arbitrary frequency dependency, J. Acoust. Soc. Am. 115 (4) (2004) 1424–1430.

12. M.G. Wismer, finite element analysis of broadband acoustic pulses through inhomogeneous media with power law attenuation, J. Acoust. Soc. Am. 120 (6) (2006) 3493–3502.

13. J.F. Kelly, R.J. McGough, Fractal Ladder Models and Power Law Wave Equations, 2009.

14. S. Holm, R. Sinkus, A unifying fractional wave equation for compressional and shear waves, J. Acoust. Soc. Am. 127 (14) (2010) 542–548.

15. A.S. Dukhin, P.J. Goetz, B. Travers, Use of ultrasound for characterizing dairy products, J. Dairy Sci. 88 (2005) 1320–1334.

16. L.A. Mandelshtam, M.A. Leontovitch, On the theory of sound absorption in fluids, J. Exp. Theor. Phys. 7 (1937) 438.

17. L.D. Landau, E.M. Lifshitz, Fluid Mechanics. Course of Theoretical Physics, vol. 6, Pergamon Press, 1975.

18. J.J. Markham, R.T. Beyer, R.B. Lindsay, Absorption of sound in fluids, Rev. Modern Phys. 23 (4) (1951) 353–407.

19. C.M. Sehgal, J.F. Greenleaf, Ultrasonic absorption and dispersion in biological media: a postulated model, J. Acoust. Soc. Am. 72 (6) (1982) 1711–1718.

20. J. Thorson, M. Biederman-Thorson, Distributed relaxation processes in sensory adaptation, Science 183 (1974) 161–172.

21. M.N. Berberan-Santos, E.N. Bodunov, B. Valeur, Luminescence decays with underlying distributions of rate constants: general properties and selected cases, in: Fluorescence of Supermolecules, Polymers, and Nanosystems, in: Springer Ser. Fluoresc., vol. 4, 2008, pp. 67–103.

22. R. Klages, G. Radons, I.M. Sokolov (Eds.), Anomalous Transport: Foundations and Applications, Wiley-VCH, 2008.

23. A.K. Jonscher, Dielectric Relaxation in Solids, Cheksea Dielectrics Press, London, 1983.

24. V.V. Pokrovskii, The Mesoscopic Theory of Polymer Dynamics, Kluwer Academic Publishers, New York, Boston, Dordrecht, London, Moscow, 2002.

25. R. Hilfer (Ed.), Applications of Fractional Calculus in Physics, World Scientific, 2000.

26. M.A. Isakovitch, General Acoustics, 1973, (in Russian).

27. V.A. Krasilnikov, V.V. Krylov, Introduction to Physical Acoustics, Nauka, Moscow, 1984.

28. Ya.B. Zel'dovich, Yu.P. Raizer, Physics of Shock Waves and High-Temperature Hydrodynamic Phenomena, vol. 2, Academic Press, New York, London, 1967.

29. G. Lebon, D. Jou, J. Casas-Vãzquez, Understanding Non-equilibrium Thermodynamics. Foundations, Applications, Frontiers, Springer-Verlag, Berlin, Heidelberg, 2008.

30. I. Prigogine, Introduction to Thermodynamics of Irreversible Processes, third ed., Interscience Publishers, A Division of J. Wiley & sons, 1955.

31. G. von Doetsch, Anleitung zum Praktischen Gebrauch der Laplace-Transformation, Mit einem Tabellenanhang von Rudolf Herschel. Oldenburg, Munchen, 1961.

32. M.V. Fedoriuk, Asymptotics. Integrals and Series. Handbook Mathematical Library, GRFML Publishers, Moscow, 1987, (in Russian).

33. R. Metzler, J. Klafter, The restaurant at the end of the random walk: recent developments in the description of anomalous transport by fractional dynamics, J. Phys. A: Math. Gen. 37 (2004) R161–R208.

34. R. Metzler, J. Klafter, Random walk's guide to anomalous diffusion: a fractional dynamics approach, Phys. Rep. 339 (2000) 1–77.

35. M. Ochmann, S. Makarov, Representation of the absorption of nonlinear waves by fractional derivatives, J. Acoust. Soc. Am. 94 (6) (1993) 3392–3399.

36. K.R. Waters, M.S Hughes, G.H. Brandenburger, G.J. Miller, On the applicability of Kramers–Kronig relations for ultrasonic attenuation obeying a frequency power law, J. Acoust. Soc. Am. 108 (2) (2000) 556–563.

37. N.V. Sushilov, R.S.C. Cobbold, Wave propagation in media whose attenuation is proportional to frequency, Wave Motion 38 (3) (2003) 207–219.

38. H.A.H. Jongen, J.M.J. Thijssen, M. van den Aarssen, W.A. Verhoef, A general model for the absorption of ultrasound by biological tissues and experimental verification, J. Acoust. Soc. Am. 79 (2) (1986) 535–540.

Index

A

Acoustic Equation, 165, 166
Acoustic Monitoring, 166
Acoustic waves, 198
AlN, 49, 50, 51, 64, 65, 66, 147, 150, 151, 152, 153, 154, 155, 156, 157, 158, 159, 160, 161, 163
anisotropy, 4, 50, 51, 52, 146, 215, 218
Average Normal Stress, 166, 173

B

Band gaps, 210
basic multicellular units, 2, 5, 40
bender element technique, 78
Biological fluids, 247
bone modeling, 2

C

cohesive material, 77, 78, 89, 95, 108, 111
Configurable, 210, 242
cortical porosity, 2, 4, 5, 7, 17, 21, 36, 41, 42

E

Electromagnetism, 180

F

fluid medium, 197, 198

G

Gas, 166
GHz range frequency, 144

H

Heterogeneous Media, 180
high coupling efficiency, 144
Hyperelastic, 210

L

laboratory tests, 78, 79, 115
Liquid, 164, 166

liquids, 144, 145, 156, 159, 166, 169, 171, 173, 175, 245, 246, 247, 249, 252

M

Monte-Carlo, 180
multiple-scattering, 126, 129, 140

N

normal mode analysis, 197, 198, 203

O

Oil Well, 180
osteoporosis, 2, 4, 26, 36, 37, 38, 40, 41, 42

P

Phononic switch, 210
photonic crystals, 125, 126, 127, 130, 135, 136, 139, 140
Pre-stress, 210

R

Relaxation losses, 247

S

seismic wave propagation velocity, 78
Soft tissue, 247
Sound absorption, 246, 247
Stokes-Type Wave Equation, 166
Stress Relaxation Time, 166

T

Tunable, 210, 240

U

ultrasound, 2, 3, 4, 7, 9, 13, 27, 36, 41, 42, 45, 46, 47, 48, 198, 247, 250, 288, 289, 292
Uncertainties, 180

V

various elastic moduli, 78
Visoelastic Media, 166

W

wave transmission, 126
Wireless Telemetry, 180